THE PROPHET
AND THE ASTRONOMER

W • W • NORTON & COMPANY
NEW YORK • LONDON

By Marcelo Gleiser

THE PROPHET
AND THE ASTRONOMER

A Scientific Journey to the End of Time

For information about permission to reproduce selections from this book, write to Permissions, W. W. Norton & Company, Inc., 500 Fifth Avenue, New York, NY 10110

Composition by Adrian Kitzinger
Manufacturing by the Haddon Craftsmen, Inc.
Book design by Rubina Yeh
Production manager: Amanda Morrison

Library of Congress Cataloging-in-Publication Data

Gleiser, Marcelo.
   [Fim da terra e do céu. English]
   The prophet and the astronomer : a scientific journey to the end of time / by Marcelo Gleiser.
        p. cm.
Includes bibliographical references and index.
   ISBN 0-393-04987-6 (hardcover)
   1. Cosmology. 2. Religion and science. 3. End of the universe. I. Title.
   QB981 .G57413 2001
   523.1—dc21   2002000538

W. W. Norton & Company, Inc., 500 Fifth Avenue, New York, N.Y. 10110
www.wwnorton.com

W. W. Norton & Company Ltd., Castle House, 75/76 Wells Street, London W1T 3QT

1 2 3 4 5 6 7 8 9 0

For Kari

Contents

Acknowledgments

I would like to thank several Dartmouth colleagues for their wisdom and generosity, for having found time out of their busy schedules to read and offer criticism to this work. At the physics and astronomy department, my thanks go to Gary Wegner, who read the complete manuscript, and to Robert Caldwell, who read parts of it and helped with some of the figures; at the religion department, to Susan Ackerman, who, in an interview, instructed me on several details of the Book of Daniel and of Zoroastrianism; and, at the English department, to Jonathan Crewe, who read the complete manuscript. I would also like to thank my Presidential Scholar, Erika Artukovic, for help with the research on apocalyptic sects, and for her attentive reading of the first three chapters of the manuscript. Luca Amendola, a colleague and dear friend from the Observatory of Rome, suggested that I looked carefully at Plato's *Timaeus* for references on cosmic catastrophism. Richard Kremer, another colleague and dear friend from the history department at Dartmouth, suggested that I look at the excellent book by Sara Schechner Genuth on comets and catastrophism in early modern astronomy.

To my editor at Norton, Angela von der Lippe, for her unfailing enthusiasm, insight, and trust in this project. To my agent, Agnes Krup, who gracefully combines efficiency and humanity in ways rarely seen.

To Ana Paula Hisayama, from Companhia das Letras in Brazil, for all her help with clearing the copyrights for the illustrations, and to Sírio Cançado, for transforming my sketches into beautiful drawings.

Writing a book is often a very selfish enterprise; the author works alone, for long hours, week or weekend alike. I thank my children—Andrew, Eric, and Tali—for being a constant source of inspiration. And I thank my wife, Kari, for her companionship, patience, and wisdom.

Prologue

Nature from one thing brings another forth,
And out of death new life is born.

— LUCRETIUS

Throughout the history of humankind, we can witness a persistent longing to understand and make peace with the passage of time. Humans, like all living things, are born, grow to maturity, procreate, and perish. But we, alone, know it. Being conscious of our own mortality is the ultimate mixed blessing. We create works of art and theories, have children, and help those in need, attempting to produce a legacy that we hope will survive our short lives. And yet, death is still the cause of much despair, of cries of injustice and tears of pain, our final defeat by nature's power to create and destroy.

Religions of the East and the West help relieve the fear of dying or the pain of losing a loved one by turning death into an event transcending physical reality. Some designate life and death as equally important parts of an eternal cycle of existence; others promise eternal life in paradise for those who abide by their precepts. In most, redemption from the specter of death does not come lightly: the transition into a new cycle of existence or into eternal life is often punctuated by horrendous cataclysms that shake the very foundations of the earth and the sky. The Druids believed in a succession of ages, each terminated when the skies fell over their heads; the Zoroastrians believed in a final judgment day, when those of high moral standards would be granted eternal life while the wicked of spirit would be destroyed by floods of fire and molten metal. The last book of the New Testament, the Revelation to John, tells of the destruction of Earth and the lower heavens by a succession of cosmic disasters, which include collisions with "blazing stars," causing the Sun to darken, the Moon to turn bloody red, and the stars to fall from the sky.

The skies, as active channels of communication between God and the people, were watched with an intense mixture of anticipation and fear, since portents of the impending end could appear at any moment. In this book, I explore religion's assimilation of cataclysmic cosmic phenomena and its influence on scientific thought through the ages, from the pre-Socratic philosophers of ancient Greece to modern-day cosmologists. Our concerns with the stability of our environment, the perpetuation of life on Earth, the reassuring sight of yet another sunrise, or the fate of the cosmos as a whole, have permeated and inspired scientific thought since its early origins. If huge asteroids did hit Earth in the past, will they hit us again? Will the Sun shine forever? Will the universe exist forever? What is our place in a changing cosmos, where stars and solar systems are constantly created and destroyed? If Earth will be destroyed, can we perpetuate our existence elsewhere in the universe? Like questions about the origin of the universe, those related to the fate of the cosmos are deeply interwoven with religious thought. Indeed, I will argue that we create a scientific world as we do a spiritual one—in order to overcome fear, to defy time, to understand our place in this world, and to justify our lives.

As a research scientist, I find that much of my motivation to study the physical nature of the universe comes from these "big questions," even if my everyday routine is filled more with long technical calculations and detailed computer programs than with metaphysical speculation. One of my goals in this book is to humanize science, to argue that our scientific ideas are very much a product of the cultural and emotional environment where they originate. It is my hope that the present exploration of apocalyptic ideas in religion and science will elucidate some of the issues that polarize the science-religion debate, as well as paint a fairly complete picture of scientific thought pertaining to the skies, from our own planet to the universe as a whole. Drawing on the Book of Daniel, the Book of Revelation, and an investigation of apocalyptic sects, art, and literature, we will examine the formation and evolution of the solar system, the extinction of the dinosaurs, Einstein's general theory of relativity, pulsars and black holes, the big bang and the inflationary universe, all the way to the latest ideas in cosmology.

Certain issues are essentially multidisciplinary in nature, and a full appreciation of their scope requires that we examine their many facets. The end of the world is one of them. The many answers humans have proposed throughout the ages are expressions of the same universal fears and expectations. It is my hope that by examining some of them we will have a better chance of prolonging our presence on this planet and, when the time comes, elsewhere in this vast universe.

This book is written for the nonspecialist, although I truly hope some of my more specialized colleagues will be attracted by its broader nature; every scientific concept introduced is explained in jargon-free language and, whenever possible, illustrated by images and analogies from everyday phenomena; it is my belief that the essence of their message can still be captured and its beauty appreciated. I hope that you the reader, after finishing this volume, will agree with me. I am confident your effort will be worthwhile, for the vistas opened by modern cosmology are truly magnificent.

I should also emphasize that it would be impossible to do complete justice to the rich history of apocalyptic ideas in religion and to the whole of astronomical sciences in one single volume. In order to keep my task (and yours!) within acceptable dimensions, I have had to exclude several deserving topics, and limit the level of detail with which some ideas have been treated. I ask my colleagues whose worthy ideas are not included here to forgive my brevity. As a partial remedy, I append a fairly long bibliography of popular science books dealing with topics of related interest. My wish is that the whole will be much bigger than the sum of its parts, and that you will share my joy in partaking in our endless quest for meaning.

PART I

"The End
Is Near!"

CHAPTER I

The Skies Are Falling

I am all-powerful Time, which destroys all things.

—*BHAGAVAD GITA*, 11.32

We are creatures bound by time. Our lives have a beginning and an end, a finite period of time, which we hasten to chop into equal segments—years, months, days, hours—in the vain hope that through this frantic counting we can somehow control its passage. But time always has the upper hand: we do grow older and we die, not knowing when, not knowing how. This well-known fact, which many people may simply brush aside as obvious, some as too disturbing, or others as just depressing, is the single most fundamental aspect of our existence. It is what gives meaning to being human.

Death gives rise to our yearning for permanence, to our constant struggle to create, be it a painting or a family, a mathematical theorem or a new recipe, something that will stay after we go, something beyond the mere memory of our existence in the minds of our friends or relatives. Memories fade from generation to generation. A few years ago, while exploring some forgotten corners of my parents' attic, I bumped into my grandparents' photo albums, packed with hundreds of yellowed photographs of their parties, relatives, friends, celebrations, and speeches, frozen moments of a time long gone. "All ghosts now," my brother Rogério quipped, in his unique sardonic style. Looking at the pictures, I wondered how much of that laughter, of that wisdom, of their many stories, is still alive in the minds of their great-grandchildren. Feeling like the missing link of a four-generation chain, I closed the albums with a deep sense of sadness, of having lost a part of my own history, now buried in photos I can't recognize.

But wait! Close to the photo albums there was a large box, made of golden cardboard. Inside I found dozens of letters my grandparents wrote to each other, to their relatives in Ukraine and Russia, to my parents when my father was studying in Boston in the early 1950s. Extremely excited, I asked Lenore Grenoble, a friend in the linguistics department—a specialist in Eastern European and Slavic languages—to help me translate the letters. But my initial excitement quickly turned into disappointment; the letters were painfully boring, full of endless details of everyday life. No deep existential message, no deep secret revealed, nothing!

It dawns on me just how selfish we the living can be. I was not trying to get to know my ancestors better; the letters and photos could have helped me with that. What I really wanted was to get to know myself better through them. After all, their history is my history, their lives part of mine, where I grew up, who my parents were. But we can't expect the past to define our future completely. Our ancestors' lives and lessons may teach and guide, but we are the ones who must make choices.

We search for meaning, for help, for companionship. We need something more than just memories and dreams: we need hope. Perhaps we can defeat the passage of time by elevating ourselves to a supernatural level, by transcending life itself. In fact, if we beat time, we may be once more with our long-departed loved ones, for it is when time's passage is suspended that life and death merge, and the dead can coexist with the living; in immortality we become godlike. Thus, we create the infinite and the eternal. Belief soothes and justifies. It inspires us all, the painter, the teacher, the scientist, the priest, the lawyer. As Saul Bellow once wrote, "We are all drawn toward the same craters of the spirit—to know what we are and what we are for, to know our purpose, to seek grace."

Our creations in pursuit of the eternal are many. In this book we will examine some of the ways we humans have attempted to defy our time-bound existence, inspired by a common link: the mix of terror and awe of the sky above us. Because of the sacred character that all cultures and religions attributed to the skies, celestial phenomena were often viewed as a manifestation of divine power, a channel through which the gods communicated with us. And the news from above could be good or bad. In many religions, signs of impending doom or punishment often come from the skies, be they celestial objects thrown by angry deities on our homes and land, blankets of thick darkness in the middle of the day, or floods that drown all but a few chosen ones. In more extreme apocalyptic texts, falling celestial objects announce the end of all terrestrial life, the end of all ends, which will bring eternal bliss for the virtuous and eternal suffering for the sinful.

Science, since its origins, has also been inspired by the sky and its mysteries. From Plato to Einstein, many of the greatest philosophers and scientists have studied the sky not just for practical purposes but in an attempt to bring the human mind closer to that of the Creator, the Great Cosmic Organizer, believing that knowledge of the natural world lifted humankind to a higher moral sphere. The pursuit of this knowledge through reason was turned into a passionate quest, worthy of a lifelong devotion. As a consequence, our accumulated scientific knowledge of celestial phenomena drove away many ancient sky-related fears and beliefs. But in spite of all this progress or, better still, because of it, many new challenges have appeared, and will keep appearing. A scientist may refer to the continuous presence of the "mysterious" as the amazing creativity of nature or—more cynically—as an expression of our own limitations as rational beings, while a religious person may call it a manifestation of the infinite nature of God. By exploring thus our ageless relationship with the sky, whether through faith or reason, or both, we will find that religion and science represent different, but complementary, facets of our struggle against time, born of the same questing spirit. This chapter begins with a survey of apocalypticism, dramatic narratives of the end of the world according to several religions, from that of the Druids to Christianity. These timeless stories provide the bare-bones imagery that will reappear throughout this book. I call them "archetypes of doom."

Celestial Messages

"Surely, our shaman knew what he was doing. Every day, he would run up to the mountain, lifting his arms toward the skies and chanting the sacred hymns of our elders, the ones that brought us health and a plentiful harvest. He knew how to talk to the gods through the language of the night, written in the bright stars and the Moon. When many stars fell, the gods were shedding diamond tears, saddened by our lack of devotion to them. If a hairy star appeared one night and stayed for many moons, something bad would happen, maybe even the death of our king. Or worse, a giant sky-serpent could eat the Sun, bringing on eternal night. We would bring fruits and animals and clothes to the top of the sacred mountain, which was the place closest to our gods, and dance and chant for as long as our shaman told us to. The mountain connected the earth and the sky."

This fictional narrative, which combines elements of many different ancient cultures without being specific to any one in particular, is a short allegorical tale of our ancestors' mysterious relationship with the skies. The gods decide how things will be, but we can plead with them and perhaps even change their minds if we know how to interact with them, speak their language. The shaman, the holy man, is the interpreter of the gods, the decipherer of their intentions as written in the skies. His actions go both ways, from the gods to the tribe and from the tribe to the gods. He possesses the knowledge needed to understand the gods, being the bridge to the unknown. As such, he is more than human, existing in a world somewhat parallel to ours, a magic reality hovering between the natural and the supernatural.

As I tried to illustrate in this short narrative, the shaman's main power arises from his "ability" to read the skies, the messages the gods send through the stars and other celestial apparitions. (The "hairy star" is supposed to represent a comet, the serpent eating the Sun, a solar eclipse.) Thus, the skies are viewed as a sort of holy scroll, which the gods use to communicate with people via the shaman. We are not quite sure exactly when this tradition started, but it is clear that divination based on the skies was already prevalent with the Babylonians well before 2000 B.C.E. Astrology was a translation of this scroll, describing how the skies influence crops and individuals, bring on a plentiful harvest or a flood, victory in wars or the death of a king. The regularity of the heavenly motions provided the basis for a coherent interpretation of the seasons, their cyclic repetition building a sense of tranquility and control. The Sun returns every day, the full Moon in about 28 days, the summer solstice in 365 days, and so on.

But the skies were not always predictable. Inexplicable celestial events did happen, and invariably brought terror to the population. This is true for the overwhelming majority of cultures throughout history. In the Old Testament, we can find some very early examples of this "sky-linked" terror in the Book of Exodus, which tells the story of the captivity of the Jews and their escape from Egypt around 1300 B.C.E., under the guidance of Moses. Two of the ten plagues sent by God to force the pharaoh into freeing the Jews are directly related to the skies. The seventh plague brought a storm of hail and thunder that spread fire and destruction throughout Egypt: "There was hail with fire flashing continually in the midst of it, such heavy hail as has never fallen in all the land of Egypt since it became a nation" (Exodus 9:24). The ninth plague brought a thick blanket of darkness that paralyzed the whole kingdom: "So Moses stretched out his hand toward heaven, and there was dense darkness in all the land of Egypt for three days" (Exodus 10:22). Leaving aside the fruitless debate of what truly happened with Egypt's weather during those times, it is clear that to the authors of Exodus the skies were symbolically linked to the punitive actions of God. This link is more important than any speculation about the physical causes of those events, in particular when we realize that they drew on a text written some six hundred years after the events took place. The meaning here is in the symbol, not in the facts.

Because of their seeming unpredictability, comets, eclipses, and meteor showers were interpreted as messages from the gods, either admonishing a people or announcing a tragic coming event; more often than not, we fear what we don't understand. Religions and countless folkloric tales tend to link such apparitions with possible disasters that follow or precede them, from a king's death to the end of the world. The first surviving reference to a comet, from a Chinese sentence from the fifteenth century B.C.E., connects its appearance with a series of murders committed by a political leader: "When Chieh executed his faithful counselors, a comet appeared."[1]

Another Chinese text, from three hundred years later, also relates a comet to the deadly actions of a political leader: "When King Wu-wang waged a punitive war against King Chou, a comet appeared with its tail pointing toward the people of Yin."[2]

Reference to comets, from about the same time, can be found as well in a few Babylonian fragments produced during the reign of Nebuchadnezzar I,

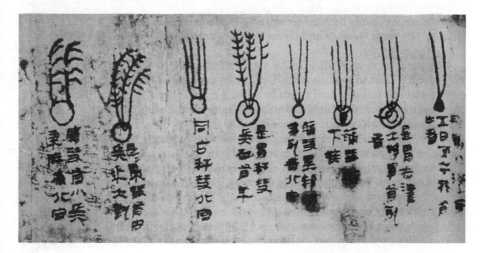

FIGURE I: *Excerpt from the first catalog of comets, ca. 300 B.C.E., the Mawangdui silk.*

"When a comet reaches the path of the Sun, Gan-ba will be diminished; an uproar will happen twice. . . ."[3] Whatever "Gan-ba" meant to the Babylonians, it clearly was not a good thing to have diminished. Mournful events on Earth were thus magically mirrored by celestial phenomena. However, these accounts of dramatic events linked to comets are fairly mild compared with their role in apocalyptic narratives, prophetic stories about the end of all ends, the time when our individual histories are blended into a single universal history. Fear of the skies and fear of the impending end have been linked for thousands of years.

Farther to the west, the fabled Druids, religious leaders of some Celtic tribes that spread across Europe starting around the sixth century B.C.E., supposedly believed the end would come from the skies. One of the most quoted Roman sources on the Druids comes from Gaius Julius Caesar (100–44 B.C.E.), the great Roman general and dictator who conquered Gaul and tried to bring the Celts, with the exception of those in Ireland, into the *pax Romana*. In book 4 of his *De Bello Gallico,* Caesar wrote, "The Druids officiate at the worship of the gods, regulate public and private sacrifices, and give ruling on all religious questions. Large numbers of young men flock to them for instruction, and they are held in great honour by their people."[4] The Druids are portrayed as the religious priesthood of the Celts, noble and proud, but also as "barbaric" and "savage," no doubt because of their central role in performing human sacrifices.

Caesar was also impressed by the Druids' belief in immortality, which he claimed explained the famed Celtic bravery in battle: "A lesson which they take particular pains to inculcate is that the soul does not perish, but after death passes from one body to another; they think that this is the best incentive to bravery, because it teaches men to disregard the terrors of death."[5] Such an attitude clearly illustrates how religious belief in the immortality of the soul serves as an antidote to our helplessness in the face of the inevitability of death. Since death marks the end of our physical existence, any attempt at overcoming it will have to invoke a supernatural reality that posits some sort of immortality, be it of the soul or of the body itself, as with the Egyptians. Modern religions, of course, are no exception.

Caesar mentions the Druids' knowledge of and interest in the skies: "They also hold long discussions about the heavenly bodies and their movements, the size of the Universe and of the Earth, the physical constitutions of the world and the powers and properties of the gods."[6]

Modern evidence seems to indicate that the Celts or the Druids did not build Stonehenge, the awe-inspiring circular monument of huge upright stones in Salisbury Plain, southern England, or others like it spread around Great

Britain, dating back to about 3000 B.C.E. Although the true identity of the builders of these megalithic structures remains a mystery, they were incorporated into Celtic culture, as astronomical observatories and sacred places for rituals. The perfect alignment of the Sun with strategically placed stones at the summer solstice shows that their positions were not accidental. The stones clocked the motions of the skies, the realm of the gods, and were thus at once useful and holy. These monuments can be thought of as a reenactment on Earth of the celestial motions, a place where a parallel magical reality was made real through ritual.

Given the Celts' attitude toward life after death, and the connection they made between the sky and magic, we might wonder what they feared the most. According to Arrian, a Greek who lived around 100 C.E., this question was asked by none other than Alexander the Great, when he met the Celts on the banks of the Danube in 335–334 B.C.E. "We fear only that the sky will fall on our heads," was the solemn answer by the Celtic chieftains.[7] The apocalyptic imagery of the Celts, as reconstructed by fragmentary evidence, envisioned the end of the world as initiated by the collapse of the sky itself, followed by fire and water swallowing up the earth and the destruction of all men. A new heaven and earth and a new race of men would then reappear. These images, found in many other invocations of the end, depict nature running wild, leaving humankind completely helpless and terrified by its overwhelming power and, more important, rendering physical life impossible. A new kind of life is necessary, a supernatural life, an existence beyond physical time.

The Bridge to Eternity

Several of the links connecting sky and doom that later played a crucial role in Jewish and Christian apocalyptic narratives originated in the ancient civilizations of the Middle East. Babylonian cataclysmic floods, which cleansed humankind of evil, as they did in the story of Noah; a paradisiacal land for deserving Egyptian souls known as the enchanted fields; or the Zoroastrian concept of the end of time at the judgment day—these are but a few examples of the belief in an afterlife, often heralded by great destruction. Without being exhaustive, we will follow some of these ideas, which presaged the grand apocalyptic visions of the Book of Daniel and the Book of Revelation.

No other culture has been so careful with its dead bodies as the Egyptian; a happy afterlife depended on the body's state of preservation, and that explains why the Egyptians took mummification techniques so seriously. Initially, only the pharaohs, being divine themselves, were entitled to a happy afterlife. However, during the breakdown of pharaonic authority (2200–2000 B.C.E.), the belief grew that an afterlife existence was the right of most, a democratization of immortality. It was in this period that one of the oldest, if not *the* oldest, detailed conception of the afterlife became very popular, that of the kingdom of Osiris, the lord of the dead.

The kingdom's location was uncertain; some accounts placed it loosely in the north, others in Syria, and still others even more remotely, in the Milky Way, "the great white Nile of the sky." The kingdom was a paradisiacal place, where the land was fertile, the grain grew high, and the chosen ones could sit under the trees, watching their slaves do all the work, while they played games, and ate and drank with their smiling friends and spouses. Of course, not everyone was granted access to such a pleasant (perhaps a bit boring?) existence: judgment had to be passed before the soul was allowed into these "enchanted fields."

As is depicted in an illustration from the Book of the Dead and many other funeral papyri, Osiris himself presided over the trial, a cross-examination of just how virtuous the soul had been (see figure 1 in insert). Anubis, the jackal-headed guardian god of the cemetery, brings the soul to Osiris, who sits stoically on his throne, flanked on one side by Isis and on the other by their sister Nephthys. The soul immediately recites all the things it didn't do:

> I knew no wrong, I did no evil thing. . . . I allowed no one to hunger. I caused no one to weep. I did not murder. I did not command to murder. . . . I did not commit adultery. . . . I did not load the weight of the balances. . . . I did not interfere with the god in his payments. I am purified four times, I am pure. . . .[8]

This negative confession is repeated forty-two times, to each of the judges representing the provinces of Egypt. Anubis then places the dead person's heart on a scale, weighing it against an ostrich feather. If the soul is light, it is allowed to enter the enchanted fields. Otherwise, retribution is swift. In some narratives, it is destroyed by a strange being, with the head of a crocodile and a body that is a hybrid of a lion and a hippopotamus. In other accounts, it is thrown into a fiery hell, where it suffers terrible punishments.

We can identify several moral notions that reappear in many other religions: the dos and don'ts of a righteous existence, the judgment of the soul by the gods, and reward in paradise or punishment and torture in a fiery hell.

Over a period of more than two thousand years, and after many cycles of invasion and domination, there was no doubt widespread religious cross-fertilization in the Middle East. But for a clear ethical distinction between good and evil, as well as several other basic elements that would be crucial parts of the Jewish, Christian, and Islamic faiths, we must consider a religion that arose in the eastern limits of the Middle East, Zoroastrianism.

Zoroaster, whose name is a Greek corruption of the ancient Iranian name Zarathustra, "the one who plows with camels," was the son of a landowner who lived in what is today the region around western Afghanistan and Iran. His birth is loosely dated to 660 B.C.E. or so, although there is great uncertainty; some sources date it much earlier, around 1000 B.C.E., while others place it as late as the first half of the sixth century B.C.E. According to tradition, he was a kind and good-natured young man who decided, at the age of twenty, to leave his parents and the wife they chose for him, in search of religious enlightenment. Greek sources indicate that, frustrated by the answers from people he met in his travels, Zoroaster spent seven years living in complete silence and isolation in a cave (extended to twenty years and eating only cheese by later Roman sources!) until the magic moment of revelation arrived, when he was thirty. In his vision, the archangel Vohu Manah (Good Thought), appearing as a figure "nine times as large as a man," invited him to leave his body behind and travel as a disembodied soul, to meet Ahura Mazda (the Wise Lord) and his court of angels. Ahura Mazda then instructed him on the doctrines and intricacies of the true religion, proclaiming Zoroaster to be his prophet.

Zoroaster's religion was a morally rigid monotheism, in which Ahura Mazda was the only true God. Mazda was the creator of the world and of man, supreme in power and in value: according to Zoroaster, although not to some of his followers, Lord Mazda willed all things into being. Because the traditions of Zoroastrianism were transmitted orally for almost eight hundred years (written versions of the Avesta, their sacred text, were first produced only during the third or fourth century C.E.), there is considerable discussion as to what the original teachings of Zoroaster were. Nevertheless, we know that he believed in the constant conflict between good and evil as a pervasive aspect of man and nature.

This conflict originated when Mazda created the world, giving freedom of choice to its human inhabitants. An individual can choose his own path, in exerting his free will. This is a brilliant solution to the problem of evil: it shifts the responsibility of goodness from God to man, without diminishing his omnipotence. Man's soul is thus the seat of a constant battle between good and evil. But people must be aware of the consequences of their choices: evil does not go unpunished. For the "Day of Final Judgment" will come, when Mazda will finally triumph over evil. As with the Celts, there was here a conception of the end of times, an eschatology rich with imagery that reappeared in later apocalyptic narratives.

According to Zoroaster, in the end of days the dead will be resurrected and, together with the living, subjected to an ordeal of fire and floods of molten metal. But while the evil ones will be burned and destroyed by the rivers of molten metal, the good and virtuous will have nothing to fear, the incandescent metal feeling like cool milk to their skins—a moral lava. In another version, each soul must be judged individually shortly after death, and must await its destiny until the end of time. The soul is to pass through the "Bridge of the Separator," which crosses over the abyss of hell, stretching outward toward paradise. During the crossing, the soul's list of deeds is read and judgment is passed. If good prevails over evil, the "pointing of the hand" (probably Mazda's) will be toward paradise; otherwise, the hand will point down, to the fiery hell below.

This version of the end of time is perhaps the first apocalyptic narrative with a single and final end, not followed by other beginnings or rebirths, as is the case in flood stories or in religions such as Hinduism, where the world is re-created in cycles. Zoroastrianism inaugurates a new era of religious thought with a linear history, the world having one beginning and one end. The end of time becomes also the end of the conflict between the two opposites, the final and eternal victory of good over evil. Immortality is achieved both by the righteous and by the sinful, the first living forever in paradise while the others are tortured forever in hell. This unforgiving nature of Zoroastrianism, with its emphasis on eternal bliss or damnation, would be taken over by Christianity's views of the end.

We create gods so that we can aspire to an abstract ideal of goodness, which we call sacred. This was as true for the followers of Zoroaster as it is for many of us today. It is our way of directing our lives and giving meaning to our actions; civilization would have been impossible without such an ideal. But this impulse toward the sacred is also our way of transcending ourselves. For after we create gods we aspire to be like them. Humans are spiritual beings, constantly yearning to establish a relationship with the mysterious, the unknown. If you are an atheist, maybe you enjoy horror movies, or other fictional connections with the supernatural. If you are an atheist who hates anything remotely related to the supernatural, maybe you worship your tennis match or the football game on Sundays, or the constant quest for perfection at sports or at work, which, as a consequence, becomes enshrined with ritualistic details. Much of religious ritual is an imitation of the actions of gods or God, an attempt to reach the sublime, a pursuit of the extraordinary, of the unreal, of the eternal. Jews rest on the Sabbath because their Lord rested on the seventh day, while Christians wash someone's feet on Maundy Thursday, the Thursday before Easter, just as Jesus washed the disciples' feet. By being like gods, we suspend the passage of time, joining the eternal. It is an ironic paradox that the very consciousness that gives us our awareness of the passage of time, and thus makes us human, needs to defeat this awareness. Our knowledge that life has a beginning and an end, that it is bracketed within a short period of time, pulls us in two opposite directions; either to accept death as the end of existence or to transcend it. This, in a nutshell, is the essence of the human predicament—to know there is an end to physical life.

Dreams of Perfection

Early Greek philosophers despised the pantheon of gods adored by the people; they were too human, suffering from too many moral frailties. How could one aspire to be like that? At around the sixth century B.C.E., a search for the One, for the fundamental reality of things, changed forever the intellectual history of the West. Thales saw everything as water, Anaximenes as air, Heraclitus as fire, and Anaximander as an abstract, intangible substance that permeated reality. Plato believed in the Demiurge, the rational cosmic artisan, as an absolute principle of pure Good, who stood above all other minor deities. His illustrious disciple Aristotle, at least in his earlier work, did away with all minor gods and deities, but needed to postulate the existence of the Prime Mover, the initiator of all motions in the cosmos, while himself being devoid of motion. (Some call him the Unmoved Mover.) Thus, we see in ancient Greece a move toward a monism, side by side with that of the Zoroastrians in Iran and that of another Semitic tribe of the Middle East, the Jews. The Greek philosophers aspired to a rational perfection, while the others aspired to a moral perfection. However, these two parallel monistic currents forked into two very different paths, and elements of Plato's and Aristotle's theological ideas were incorporated into a monotheistic religion only with the advent of Christianity, a few centuries later.

Aside from its influence on Christian theology, Greek monistic thought is of great importance to the development of modern science, in particular after the rediscovery of Greek texts in Europe during the Renaissance, courtesy of the Arabs. Modern physical science is built upon an abstract structure based on fundamental laws that dictate the behavior of natural systems. These laws are discovered both by inspection—that is, by a painstaking comparison of data with mathematical models—and by deduction, which starts with a theoretical foundation and then makes specific predictions subsequently contrasted and compared with observational data. One then arrives at a model, resting on a specific set of fundamental laws, that describes a host of natural phenomena. This procedure is self-correcting and self-improving: models, and the laws they are based upon, are continually scrutinized by the scientific community, the goal being the explicit understanding of their limitations and weaknesses. Once the scientific community arrives at a consensus, the model is accepted and used, within its limits of validity, to describe as many phenomena as it possibly can.

For example, the three laws of motion, as stated by Isaac Newton in his 1687 book *Mathematical Principles of Natural Philosophy*, accurately describe motions at speeds familiar in our everyday life, such as those of cars, airplanes, and spaceships. However, they fail to describe motions close to the speed of light (186,000 miles per second), or on very small spatial scales, such as those of molecules or atoms. Newton's laws of motion, and any other law of nature, have been shown to be valid since the dawn of time (or almost) and anywhere in the universe; as such, they are said to be universal laws. Science, at its most fundamental level, can be thought of as the pursuit of these laws, which describe different levels of physical reality, from the microscopic to the macroscopic. Being a true heir of Greek monism, science aims at an abstract rational perfection, although its practitioners know that true perfection is unattainable; a continuous process of self-refinement propels science forward. To a scientist, nothing could be more central than to partake in this ritual of discovery.

Hence, a clear parallel between the goals of science and those of religion emerges from this discussion. Both aim at something beyond the ordinary, an abstract ideal of perfection, which transcends the human dimension. And yet, science cannot provide the emotional comfort that religion provides to so many people, while religion cannot provide the rational dimension that science provides. The fact that they have something in common does not mean that they have the same purpose. You don't ask a physicist or a chemist for solace when someone you love dies, at least not as a professional, and you don't (or shouldn't) ask a rabbi or a priest for explanations of quantum mechanics (unless, of course, the rabbi or the priest is a physicist, always a possibility). We need both pursuits, and closing our door to either leads only to confusion. Just as it is absurd to say that Earth is six thousand years old, it also does not make sense to say that science has all the answers, or even that it is capable of having all the answers. Some questions simply don't belong to science.

Visions of Doom and Redemption

The many apocalyptic themes and images found in ancient Babylonian, Egyptian, and Zoroastrian beliefs crystallize in the Book of Daniel, written when the Jews were suffering terrible religious persecution from the Syrians during the second century B.C.E. Considered to be the first truly apocalyptic text, the Book of Daniel powerfully describes the impending end as seen by the prophet in visions charged with sociopolitical symbolism. It also offers a clear link between doom and the disruption of the sky and its natural order, which would reappear in several Christian apocalyptic texts. As we will see, some of the imagery found in the Book of Daniel was used during the seventeenth century by preachers and natural philosophers alike, as they invoked the fear and the science of the "End of Time." Before I discuss the apocalyptic imagery of the text itself and its celestial links, it is important to survey the historical background that prompted its writing.

In 332 B.C.E., Alexander the Great defeated Darius III of Persia, initiating Greek expansionism toward Asia and Africa. Surrounded by Hellenic culture, the Jews of Palestine and the Diaspora responded in different ways; while some were excited by Greek theater, philosophy, sports, and poetry, others were disturbed by this foreign influence. The Greeks were quite tolerant of Jewish religion; some even participated in their religious rites, placing the Jewish god Yahweh (or Iao, as they called him) side by side with Zeus and Dionysus in their list of high deities. The Jewish ritual, with its absence of statues and emphasis on oratory, must have seemed to them a variation on a philosophical debate, the synagogue as a substitute for the school of philosophy.

Alexander's early death in 323 B.C.E. led to one hundred years of invasions by rulers of his prior provinces, the Seleucids (Syrians) and the Ptolemies (Egyptians). In 198 B.C.E., the Seleucids finally won control and things were peaceful for a while, until King Antiochus IV Epiphanes, who reigned from 175 to 164 B.C.E., decided to unleash a policy of religious intolerance, imposing the cult of Zeus on the bewildered Jews. In December 167, Antiochus erected an altar to Zeus in the Temple of Jerusalem, ordering different sacrifices to be performed there, such as the killing of pigs, to the horror of the Jews. The turning point came when an elder Jewish priest named Mattathias was forced by a Syrian commissioner to participate in a sacrifice for Zeus in the village of Modein. Mattathias ended up killing the commissioner, unleashing a strong revolt against the Syrian domination.

The prominent figure in this remarkable story was one of Mattathias's five sons, Judas Maccabeus, whose bands of Jewish soldiers, to the complete astonishment of the Syrians, defeated four of their armies and forced a fifth to retreat. In 165 B.C.E., Judas managed to reconquer Jerusalem, cleansing the Temple from the Hellenic "abominations" and rescuing Judaism in Palestine. There followed a period of peace and independence that lasted until 63 B.C.E., when a civil war between different factions among the Jews, the Sadducees and the Pharisees, ushered in the Roman dominance of Palestine.

It was during the reign of Antiochus IV Epiphanes, a period of great chaos and pain to the Jewish people, that the Book of Daniel was written, probably between 167 and 164 B.C.E. In it we find the first of several Jewish apocalyptic texts that became quite popular between 150 B.C.E. and 100 C.E. Although it is difficult to trace the exact roots of the rich and fantastic imagery used in the Book of Daniel, it has been argued that many of the elements of Jewish apocalypticism display Zoroastrian influences.[9]

The Book of Daniel, it is clear, was written as a morale booster to the Jews, an antidote to their sufferings under a despotic and castrating foreign ruler. It is a book of hope, of final triumph for those who remain loyal to their faith and to their God through difficult times. Daniel appears as a special person, a visionary capable of deciphering dreams and of surviving in a den of lions for a whole night. His powers come from his faith, from his devotion to the one and only God. The first six chapters are set in the past, during the Babylonian rule, and demonstrate how tyrants who place themselves above God are destroyed by their own vanity, a lesson also taught in Exodus: no one can confront the Jewish God.

After the first six chapters established Daniel's powers and his intimate alliance with God, the final six turn to the apocalyptic. In chapter 7, Daniel sees God, "the Ancient One took his throne, his clothing was white as snow, and the hair of his head like pure wool; his throne was fiery flames, and its wheels were burning fire" (7:9). God was surrounded by fire and by a court of thousands who served and stood by him. "The court sat in judgment, and the books were opened" (7:10). Daniel then sees

> one like a human being coming with the clouds of heaven. . . . To him was given dominion and glory and kingship, that all peoples, nations, and languages should serve him. His dominion is an everlasting dominion that shall not pass away, and his kingship is one that shall never be destroyed. (7:13–14)

We can easily discern elements of the apocalyptic narratives of the Zoroastrians, with God and his court passing judgment and a messianic figure inaugurating a new age of peace and prosperity for the people of Israel.

In chapter 8, Daniel relates another vision, of a fierce battle between the ram (Medes and Persians) and the goat (Greek empire). After the goat destroys the ram, four horns sprout from its head (the dissolution of Alexander's empire) and one of them (the reign of Antiochus IV Epiphanes) "grew as high as the host of heaven. *It threw down to the earth some of the host and some of the stars, and trampled on them*" (Daniel 8:10, my italics). This image, linking the disruption of political and cosmic order, is a recurrent one in apocalyptic narratives. Stars and other celestial objects tend to come down from the skies in times of trouble. The vision ends with the archangel Gabriel reassuring Daniel that this evil king will "be broken, and not by human hands." The book concludes with the prediction of the fall of Antiochus's kingdom and the final redemption of the people of Israel. This revelation comes to Daniel from a supernatural narrator, with "body of beryl, face like lightning, eyes like flaming torches, arms and legs like the gleam of burnished bronze, and the sound of his words like the roar of a multitude" (10:6). This narrator tells Daniel of the arrival of

> Michael, the great prince, the protector of your people. . . . There shall be a time of anguish, such as has never occurred since nations first came into existence. But at that time your people shall be delivered, everyone who is found written in the book. Many of those who sleep in the dust of the earth shall awake, some to everlasting life, and some to shame and everlasting contempt. (12:1–2)

Things will thus get worse before they can get better. The day of judgment will also be the day of the resurrection, when the dead will arise and join the living in either eternal bliss or eternal damnation (see figure 2 in insert). The story has a tragic twist: peace and redemption can come to the Jews only through a supernatural action, when God himself will take matters into his hands and destroy the enemies of Israel. As the text makes clear, this can happen only at the end of history, the end of time as we know it, when the dead and the living can coexist. There is no salvation in "real" time.

Daniel is a bridge to divine knowledge, the interpreter of things to come. If the visions are fantastic and otherworldly, it is because they must be so; no religious imagery based on concrete reality would have such an effect on people's imagination. After all, we are speaking of events with an epic moral scope, the perpetual end to suffering, the final vindication of goodness. It is the end of human existence as we know it, the beginning of the reunion of man with God, as it had been long ago, when God and man walked side by side, before the expulsion from the Garden of Eden. The apocalyptic narrative foretells the transformation of the faithful and virtuous into godlike creatures and the devious and perverse into devil-like ones. It must have been very hard to resist its alluring power; to many people, it still is. The narrative invited people to participate in a grand scheme laid down by God himself, to endure evil times as they were short, and never to give up hope that vindication was at hand and that it would be definitive. In fact, put into context, people's sufferings were to be viewed as a necessary part of the whole thing, giving it meaning, placing it within the unfolding of the events to come. As Saint Lactantius wrote a few centuries later, "virtue is the ability to endure hard things." The more you can endure suffering, the more virtuous you are, and the more wonderful will be your rewards in the hereafter. What a great recipe for social paralysis! Endure your suffering with virtue and grace, have faith, and God will take care of the rest for you. It is no wonder that these concepts were elevated to absurd levels during the Middle Ages, when pestilence, despair, and lack of leadership reached absurd levels. The darker the times, the more people need hope, the more they need to have faith in their redemption. The darker the times, the more vulnerable people are, the more readily they fall prey to blind fanaticism—a tendency that should not be overlooked in our own day.

Revelations of the End

The apocalyptic ideas put forward in the Book of Daniel percolated across several other Jewish texts and also inspired early Christianity, when it too was a persecuted sect. The Roman occupation of Palestine, which started in 63 B.C.E., ended tragically in 70 C.E. with the destruction of the Second Temple in Jerusalem and a new exile of the Jews. The reversal of what was an unusually tolerant religious policy came after a radical Jewish resistance group known as the Zealots openly opposed the Roman dominance, holding the Roman armies back for four years. The Zealot resistance was eventually overthrown by the Roman emperor Vespasian, the Temple was burned to the ground, and, to complete the humiliation, Jerusalem was renamed Aelia Capitolina. By that time, though, another religion was quickly spreading through Palestine and beyond, based on the teachings of a new prophet, Jesus of Nazareth.

The turning point in Jesus' life came when he was about thirty. (There is an uncertainty of a few years because of contradictory dating in the Gospels.) An ascetic known as John the Baptist, probably from the highly spiritual Jewish sect known as Essene, was summoning the masses: "Repent! For the Kingdom of Heaven has come near!" (Matthew 3:2). John was convinced that the end was imminent and that only those who repented their sins would stand a chance when judgment came. There was great expectation among the Jews that the Messiah was coming, and that he was a normal man, a descendant of King David, founder of Jerusalem. Receptivity is key to the success of any religion. People from all over Palestine came to listen to John. Those who chose to follow him were then bathed in the river Jordan, to wash away their sins, an Essene purification rite. John instructed his followers to share their clothes and food with those in need and to be generous to their neighbors. It was after his baptism that Jesus had his vision: "[The] heavens were opened to him and he saw the Spirit of God descending like a dove and alighting on him. And a voice from heaven said, 'This is my Son, my beloved Son, with whom I am well pleased'" (Matthew 3:16–17). John's apocalyptic message lies at the very heart of Christianity and was preached by Jesus himself, although with some important changes. For Jesus, redemption was open to all; the kingdom of the Father was for all righteous men, not only Jews.

We should keep in mind that the first of the Gospels, that of Mark, was written some seventy years after Jesus' death, around 30 c.e., and that distortions of deeds and sayings, as well as later editorial changes, must have taken place. It is not unlikely that the conflict between Jesus and Jews was duly exaggerated in later texts, for propagandistic reasons. This would certainly have widened the split between Judaism and a new emerging religion, attracting those unhappy with the Jewish religious leadership and its quarreling factions.

All three synoptic Gospels (from *synoptikos*, "seen together," Matthew, Mark, and Luke) include Jesus' description of the arrival of the "End of the Age," when the Messiah will come down from heaven in a cloud. In all three, the signs of the end are clearly cosmic signs, doom coming from the skies: "Immediately after the suffering of those days the sun will be darkened, and the moon will not give its light; the stars will fall from heaven, and the powers of heaven will be shaken" (Matthew 24:29). Essentially the same descriptions can be found in Mark 13:24–25 and in Luke 21:25, although Luke is less explicit about specific cosmic events. The apocalyptic narrative from the Book of Daniel and other intertestamental texts announcing the end of time and the arrival of the Messiah is combined, at the very birth of Christianity, with signs of doom coming from the skies. These prophecies create a state of constant anxiety with regard to cosmic events; every shooting star, every eclipse, every comet or unexpected celestial event may be interpreted as being part of the doomsday prophecy, the harbinger of the end to come. As a result, people look up to the sky with a mixture of hope and terror, commensurate with the level of social despair; in quiet times (no doubt quite rare!), the skies are usually less threatening. For those with a guilty conscience, celestial events give life to the saying "God is watching you," warning that your sins will not pass unpunished. For the suffering pious, the skies offer the hope of a better afterlife and the celestial signs are eagerly awaited. In either case, God's message of doom and hope is written in the stars.

Eschatological views made appearances in several of the Apostolic letters (for example, in 2 Peter 3:3–13 and in 1 John 2:18), but reached an absolute climax in the Revelation to John, the last book in the New Testament canon. (In fact, its inclusion in the New Testament was a very controversial issue in the early church.) According to historians of religion, Revelation was probably written around 95 C.E. by a prophet named John, the leader of a Christian circle in Ephesus, on the western coast of Turkey. After the fall of Jerusalem in 70 C.E., the teachings of Jesus quickly spread across the western and northern Mediterranean coast all the way to Rome, through the courageous actions of Apostles such as Paul, Barnabas, and many others. These departures from the "old faith," that is, the respected Judaism, were looked upon as trouble by the Romans, a manifestation of impiety. For them, to claim that Plato, Aristotle, and Alexander the Great were sons of a god was quite acceptable. But to claim that a Jew who died a terrible death in a faraway land was not only the son of a god, but *the* Son of God, was sheer lunacy. As a result, there was widespread persecution and severe punishment of Christians at least since 64 C.E., during Nero's rule, which included exile or death. "The Christians to the lions!" was a common cry. Quite possibly, John's prominence saved him from being fed to the lions, and he was banished to the island of Patmos, near Ephesus, his birthplace.

Revelation was written as a response to the Roman anti-Christian sentiment and its threat to the survival of the church. Like previous apocalyptic texts, it encourages the faithful to endure the persecutions and the sufferings, exhorting martyrdom while promising the arrival of the end, when justice will prevail and goodness be vindicated. However, the language and symbols used by John, sometimes quite beautiful and poetic and sometimes absolutely terrifying, are by far the most powerful found in apocalyptic narratives. The sublime is mixed with the horrific, with the clear intent of scaring and moving people, to awaken them to the Christian faith—if, that is, they are interested in eternal salvation. The document is brilliant propaganda, still holding its power as the banner of many apocalyptic sects today, sometimes with tragic consequences. After congratulating some churches for their work and censuring others for slackness, John proceeds to relate a series of visions in three cycles, each with a set of seven symbols (specifically, the seven seals of the scroll, chapters 6–7; the seven angels blowing their trumpets, chapters 8–10; the seven bowls of God's wrath, chapters 15–16). The narrative is interspersed with visions of God on his celestial throne, of horrible beasts of all sorts rampaging the earth, of violent cosmic cataclysms, of the battle between the archangel Michael and the dragon (Satan), and of the victory of Christ over the Antichrist, inaugurating Christ's thousand-year kingdom on Earth (the doctrine known as chiliasm). After this period, Satan is again liberated from the "bottomless pit," and the final battle between good and evil takes place, marking the end of time and the appearance of a new heaven and a new earth, the New Jerusalem.

Every one of the visions is punctuated by cosmic cataclysms, signifying not only the absolute power of God over nature but also his use of nature's power to express his wrath. In chapter 6:12–14, when the Lamb (Jesus) opened the sixth seal, "there came a great earthquake; the sun became black as sackcloth, the full moon became like blood, and the stars of the sky fell to the earth as the fig tree drops its winter fruit when shaken by a gale" (see figure 3 in insert). After the seventh seal was opened, an angel "took the censer [he was holding in front of God's throne] and filled it with fire from the altar and threw it on the earth; and there were peals of thunder, rumblings, flashes of lightning, and an earthquake" (Revelation 8:5). The image of a censer filled with fire being thrown on Earth is strongly evocative of a comet or a large shooting star. The narrative then shifts from the seven seals to the seven trumpets, all blown by angels, one at a time. Each sounding of a trumpet brings a more horrific disaster on Earth, in an attempt to coerce the sinful to repent. When "the third angel blew his trumpet, a great star fell from heaven, blazing like a torch, and it fell on a third of the rivers and on the springs of water" (8:10). Here we have another description inspired by a falling celestial object causing terrible damage on impact. All cosmic events associated with apocalyptic narratives were unpredictable; in the cultures that generated them, there was a natural division of celestial phenomena into the cyclic, the ones they knew would return regularly, and the random, the ones that appeared suddenly. Comets are somewhat peculiar here, since many of them do reappear cyclically, albeit usually with quite long orbital periods. The most famous of our cometary visitors, Halley's comet, has a period of approximately seventy-six years (see figure 4 in insert). To predict the return of a comet without knowledge of the workings of gravity requires the existence of a sustained astronomical culture for prolonged periods of time, something quite rare in antiquity. With the exception of the Bantu-Kavirondo people of Africa, the idea that comets do in fact reappear at regular intervals had to wait until the early eighteenth century, when Edmond Halley applied Isaac Newton's theory of universal gravitation to cometary motion. And, as we will see, even then cometary science was deeply enmeshed in religious doomsday rhetoric, apocalyptic fears, and apocalyptic science interwoven by popular culture.

The nature of what constitutes a terrifying celestial apparition changed over the ages; the more science could quantitatively describe the skies, the less terrifying several apparitions became. Thus, eclipses are now predictable to the second, comets are observed sometimes years before they become visible with the naked eye (or even when they aren't), and meteors, bits of rock that fall onto the earth and are ignited by friction with the atmosphere, a.k.a. shooting stars, if not predictable, inspire more wonder than fear. Unless, of course, they are big bits of rock, too big to be vaporized by the atmosphere and end up hitting the ground, in which case they are called meteorites. These "random" celestial events invariably found a place in religious apocalyptic narratives. Since the skies were controlled by the gods or by God, they had to be conveying a divine message. And since God usually did not interfere with world affairs unless things were going badly, these messages were, for the most part, signs of anger or upcoming trouble. This notion carved itself deeply into the psyche of the Western world. Even though the final intent of apocalyptic narratives was the promise of justice and peace to the faithful, the skies were tinged with fear. The passage into eternal life or damnation was to be heralded by celestial chaos: hell comes from the heavens.

CHAPTER 2

Heaven's Alarm to the World

What if this present were the world's last night?

—JOHN DONNE (1572–1631)

The title of this chapter comes from a 1681 sermon by Increase Mather (1639–1723), president of Harvard College and "teacher of a church in Boston in New England." Increase was the son of the Puritan Richard Mather, who sailed for America in 1635 to preach what he could not at home, and father of the famous missionary Cotton Mather. He firmly believed that unexpected cosmic events, and the calamities preceding and following them, were messages from a very unhappy God to the crowd of sinners down below. In fact, the full title of the sermon reads,

Heavens Alarm to the World.
Or
A SERMON
Wherein is shewed,
That Fearful Sights and Signs in Heaven
Are the Presages of great Calamities at hand.[1]

The message of the sermon was clear: to urge members of the congregation to repent their sins, because cosmic signs of the end were revealing themselves as prophesied in several biblical texts, from Jeremiah to Joel to Revelation, which

he quoted profusely amid his inflamed rhetoric. A believer in the literal meaning of the Bible's words, Mather saw the skies as a stage where God enacted his
wrath. Two years after his 1681 sermon, he published his *Cometography*, an
encyclopedic survey of all recorded (and some invented) comets, "from the
Beginning of the World unto this present year, 1683,"[2] and their associated
calamities. Thus, in 984 c.e., five years before the passage by Halley's comet, "a
Blazing Star was seen; an Earthquake, Wars, Plague, Famin[e]s followed, the
Emperor and the Pope died."[3] In 1005 c.e.,

> a Blazing Star horrible to behold was seen flaming in the Heaven in the
> Spring time for the space of thirteen days and then again in October.
> There followed the most fearful Plague continuing for three years. In
> some places whilest the living were burying the dead, they drew their
> last breath, and were thrown into the Grave with those whom they
> intended to leave there.[4]

Countless variations on these basic themes may be found in the Reverend
Mather's survey. The comet of 1680, the one that no doubt inspired his sermon,
appeared, as Donald K. Yeomans remarked in his chronological history of
comets, during "the zenith of cometary superstition": "in Germany alone there
were nearly 100 tracts published. Of these, only four were written to quiet
superstitious fears."[5] A medal struck in Germany to commemorate the appearance of the comet carried this inscription: "The star threatens evil things: trust
in God who will turn them to good."[6] The stern Reverend Mather was not
alone in his beliefs. In the late seventeenth century, comets provoked the same
fears as two thousand years before; they were taken to be signs of impending
doom, of disease and death of important rulers, of wars and famines, of chaotic
natural events beyond anyone's control. Thus, in the minds of most people,
comets were not accidentally correlated with these tragic events: they were sent
down from heaven as messengers of an angry God. They were religious entities, performing a function prescribed by sacred texts. In other words, comets
were not natural, but supernatural, phenomena.

Inspiring as it was to preachers of doom, the comet of 1680 was destined to
become a turning point. In 1687, Isaac Newton established a quantitative
method devised for computing the motions of celestial objects, following observations of this very comet. In the next chapters, we will discuss the scientific
importance and repercussions of Newton's theory of gravitation to cometary
science and to astronomy in general. But now I want to explore a more obscure
topic, namely, Newton's motivations for studying the physics of the heavens.

Newton spent a large amount of time trying to date biblical events. Not just dating them, but, as one might expect from Newton, precisely dating them. He wanted to quantify prophecy, fitting it within history. In his *Observations upon the Prophesies of Daniel and the Apocalypse of St. John*, published posthumously in 1733, he wrote that prophecies offer "convincing argument that the world is ruled by providence."[7] That is, with the proper quantitative approach—the science he was inventing—it was possible to actually interpret the Bible as a calendar of events, those in the past and those still to come, as revealed by God to the prophets. The climax of this linear unfolding of prophecies was, if not as an exact prediction at least as a certainty, the eventual arrival of the judgment day.

I confess my surprise when I first encountered this side of Newton's work. After all, how could the man who seems the embodiment of rationality and cool scientific judgment be involved in pursuits that so clearly mix science with superstition? The answer is that, for Newton, as for many (but not all) of his contemporaries, there was no clear-cut distinction between science and religion. In his view, the scientist's (or, to use the more appropriate term for the times, the natural philosopher's) search for a quantitative description of natural phenomena was part of a grander quest, that of deciphering God's plan, or mind: the scientist was a decoder of God's writing. No wonder Newton has been called "the last of the magicians."[8] And yet, Newton was careful not to publicize his work on biblical chronology or alchemy. Was he afraid of exposing himself to the ridicule of some of his less mystically minded colleagues? Quite possibly. It is certainly well-known that Newton had a great aversion to academic disputes, partly because of his character and partly because of some painful experiences early in his career. I can only imagine the enormous satisfaction and moral encouragement Newton extracted from his secret labors, no doubt believing that when the end came his efforts would be more than vindicated.

Although over fifteen centuries had passed since John of Patmos received and duly recorded his Revelation, in Newton's time belief in the end was as strong as ever. Extremely resilient, these beliefs did not disappear with the onset of the so-called Enlightenment of the eighteenth century, when rationalism reigned supreme, although many people considered them foolish old superstitions. And they certainly haven't disappeared today. Just think of the proliferation of apocalyptic sects throughout the world, sometimes with tragic consequences, as in the 1993 incident between the Branch Davidians and the U.S. government in Waco, Texas, or in the 1997 suicide of thirty-nine members of the Heaven's Gate order in Southern California. Or the all-pervasive apocalyptic imagery in the media, referring to real or fictional threats such as a nuclear holocaust, alien invasions, collisions with asteroids and comets, or global warming. Not to mention generalized end-of-civilization-as-we-know-it social panics, such as the so-called Y2K computer threat, the "millennium bug." In this instance, countless books were written and web sites selling survival kits created, in what amounted to an amazingly profitable techno-farce. But skeptic that I am, I backed up all my computers in late December and took my family to Rio for the New Year, where at least we would all be warm.

Without a doubt, eschatological visions and apocalyptic fears are very much a part of our technological world, of our "age of science," the ancient rhetoric merely being recast into modern parlance. Floods may now come from global warming, pestilence from a vicious biological warfare, the poisoning of the soil, air, and water from industrial pollution. Some observers may even say that the very development of science brought doomsday closer—that now, for the first time in our history, we don't have to wait for God to decree the end: we can do that ourselves, since we hold the key to our collective oblivion. What was once found in sacred texts is now a palpable reality, forcing us to reevaluate how to deal with our collective survival or, more to the point, how to ensure it. But my more hopeful view is that by examining the reasons for the persistence of apocalyptic thought in our culture, we will stand a better chance of relegating it to just that, culture—and, as a result, have a better understanding of our own humanity.

The Importance of Being Vague

Vagueness can be an extremely powerful propagandist weapon; by insinuating without committing, by suggesting without defining, by creating fantastic images that inspire and horrify without offering a clear interpretation of their meaning, you are certain to attract the attention of many people. If your message ensures there is a reward for those who follow you and your ideas, some sort of relief to the tribulations of life, your success is practically guaranteed. Better yet, because you leave so much room for interpretation, extracting the detailed meaning (or meanings) of your message can turn into an obsession for many generations to come. So it was with the Book of Revelation. The stakes were high, eternal salvation or damnation. There was much disagreement among the early church fathers over how to distill John's visions. Some of this confusion was due to exogenous factors, such as natural calamities and invasions by German tribes and later by Muslim armies, while some was due to endogenous factors, such as schisms and power struggles within the church. In times of great distress, a literal reading of Revelation was favored, whereas quieter times nurtured a more abstract approach. These disputations were fed by the structure and style of the text, which cleverly lent itself to so many different readings. Its vagueness was its virtue.

One of the first Christian apocalyptic sects inspired by the powerful rhetoric of Revelation appeared in 156 C.E. in Anatolia, Turkey. A man named Montanus declared himself the incarnation of the Holy Ghost, who, according to the Fourth Gospel, was supposed to be the voice of things to come. Montanus was soon joined by a band of ecstatics, who fervently believed in the imminent coming of the New Kingdom; the heavenly Jerusalem was about to descend from the skies. The extent to which people believed in such visions and prophecies should not be underestimated: Tertullian of Carthage (ca. 160–after 220), the greatest theologian in the West at the time, after joining the movement declared that in Judea a walled city, no doubt the New Jerusalem, had been seen floating in the sky every morning for forty days![9]

The movement inspired a militancy that was to be characteristic of many future millenarian sects, movements that believed in the imminent arrival of the Messiah. It served a deep emotional function for those who joined, transforming the suffering and the persecuted into agents of change, into key players in God's grand scheme for humanity, the apocalyptic drama. It made Christians into the chosen ones, a role previously reserved for the Jews. It also created a kind of apocalyptic blindness, where all measure of reality was lost to the grander goal of the mission. The natural and the supernatural were completely intertwined: if holy cities could descend from the skies to fulfill prophecies, why not comets and other celestial phenomena?

It was Saint Augustine (354–430), the Algerian-born bishop of Hippo, who defined the standards for interpreting apocalyptic and other biblical texts. In *The City of God*, Augustine argued that Revelation was to be read allegorically, that the struggle between good and evil was already present in people's lives, and that the church of Christ *was* the Kingdom of God on Earth; it had arrived already. There was no use in trying to decipher God's logic in human terms, by predicting when and how the last judgment would arrive. Dating the apocalypse was a superfluous task, leading necessarily to ambiguity.

This is not to say that Augustine did not believe in the last judgment; he surely did and urged people "not to be slow to turn to the Lord . . . for His wrath shall come when you know not."[10] The believer is to live in a state of constant apocalyptic angst, where every action counts toward the last judgment, which may be just around the corner or millennia away. It was an extremely clever way of securing the meaning of the message from Revelation without discrediting the church in case prophecies failed to materialize.

Augustine was also concerned with the interpretation of the world's end, which seemed to be described differently in Old and New Testament books. In particular, he was distressed by some Hebrew texts, which predicted the end of *both* heaven and Earth, as in Psalms 102: 25–26: "Long ago You laid the foundation of the earth, and the heavens are the work of your hands. They will perish, but You will endure." To counter this view of utter destruction of the cosmic framework, Augustine offered the tamer Christian view, in which only the world and the lower part of the heavens perish, thus preserving the sanctity and eternity of the upper heavens, the realm of God and the righteous. Otherwise, where would the souls go when the end came? He strengthens his argument by quoting from Matthew 24:29, that "the stars will fall from heaven, and the powers of heaven will be shaken," as meaning that if the stars fall *from* heaven, heaven remains up there: "This expression, then, is either figurative, as is more credible, or this phenomenon will take place in this lowest heaven."[11] He proceeds to support this interpretation by quoting from Virgil's epic poem *The Aeneid*, finished around 19 B.C.E., which tells of the fall of Troy. Aeneas's father, Anchises, is refusing his son's urging to leave his palace and the crumbling Troy before it is too late. To help him out of his quandary, Anchises asks for a sign from the gods and is promptly granted quite a display:

> The old man had hardly spoken when from our left came
> A sudden crash of thunder, and a shooting star slid down
> The sky's dark face, drawing a trail of light behind it.
> We watched the star as it glided high over the palace's roof,
> And blazing a path, buried its brightness deep in the woods of
> Ida; when it was gone, it left in its wake a long furrow
> Of light, and a sulphurous smoke spread widely over the terrain.
> That did convince my father.[12]

Virgil and Saint Augustine were basing their ideas on the Aristotelian view that meteors, as well as comets, belonged to the lower heavens, being in fact equated with atmospheric phenomena, as in *meteor*ology. Atmospheric, but not natural phenomena! To both Virgil and Augustine, the connection between these celestial apparitions and natural phenomena was never there in the first place, because they were God-sent entities, serving a well-defined supernatural purpose. Stars and comets were playthings in the hands of the gods, plucked from the lower heavens as dictated by Aristotle's cosmic scheme, where change was relegated to the sublunary sphere, with the higher spheres being perfect and immutable.

The boundaries between the magical and the real were blurred, creating great fear and confusion; all cosmic phenomena were attributed to supernatural causes. In a world where evil was a palpable presence and temptation was everywhere, only the church could offer guidance and support. Every aspect of life (and death) revolved around religion and superstition.

Wars of Redemption

The eagerness to believe runs deep in the minds of the discontent. In spite of Saint Augustine, apocalyptic expectations were very much a part of people's lives during the early Middle Ages. Even after Constantine the Great, ruler of the eastern wing of the Roman Empire, known as the Byzantine Empire, embraced Christianity in 324 C.E., waves of invasions by Germanic and Asian tribes, as well as internal corruption, kept eroding Rome's former glory and frustrated the attempts to restore it. Christian poets of the late fourth century, such as Prudentius (348–after 405), would express their despair in gloomy verses chanting the crumbling cosmic order,

> One day the heavens will be rolled up as a book,
> The sun's revolving orb will fall upon the earth,
> The sphere that regulates the months will crash in ruin.[13]

This alternation of stable and unstable periods ended in the mid-sixth century. In the 570s the bubonic plague made its first entry into the West, following a trail of devastation that started in Asia Minor and ran through the Middle East. Renewed Germanic invasions, this time under the half-pagan Lombards, brought horror to the northern region of the Italian peninsula. The urban populations shrank, power was decentralized, and isolationism became the rule.

Self-proclaimed messiahs appeared everywhere, sometimes amassing huge followings. Most stories have a similar pattern; a man or, occasionally, a woman has some sort of epiphany, and escapes into the wilderness for a period of meditation. The person emerges from the woods transformed, enlightened, and starts calling himself the new Messiah, or at least God's messenger. People flock around this person, some being thrown into a state of religious frenzy by the prophet's preaching. Then the miracles start: the chronically ill are healed just by touching the holy man, the blind see again, the deaf hear, and the mute speak, while strange celestial phenomena confirm the prophet's mission. The whole action is a ritual reenactment of what happened when Christ was alive, transporting people to a holy time, promising them justice everlasting. Hopes soar that this is indeed the Second Coming. But time passes and the Second Coming doesn't come. In some cases, people's patience ran out and the prophet ended up exiled or oftentimes killed. Or the movement assumed militant dimensions, becoming a revolutionary force within the poor and the socially displaced, threatening the establishment. The movement would eventually be crushed by the church or nobility or by an alliance of the two, its leaders slaughtered and their mangled bodies (or parts of them) displayed publicly as a warning against future insurrections. But apocalyptic hopes would not go away this easily. Faith is fed not by a rational search for truth but by a passionate belief in its divine revelation. In the gruesome early Middle Ages, the supernatural reality offered by the church was the only glimmer of hope. People's perception of reality was indistinguishable from the fantastic.

As the historian Norman Cohn argues in *The Pursuit of the Millennium,* the more revolutionary aspects of messianic movements came later, around the eleventh century, as a result of the development of commerce and the consequent increase in the urban concentration of wealth. Up until then, European society was mostly agrarian, obeying age-old rules, which bound the lords and their serfs in a covenant of mutual dependence and security. These agrarian nuclear societies enjoyed some precarious but comforting stability, and not much change was expected or hoped for. With the development of trade, mostly because of the emergence of the textile industry, the peasantry confronted a new reality, where wealth was possible to those either skilled enough or cunning enough to succeed in the marketplace. Temptation settled in, and many migrated to the new townships, shaking off their vows of servitude in search of newfound dreams, which, however, were rarely fulfilled. The individual felt left behind, a useless appendage to society. When all seemed lost, there appeared a savior, promising a new life to those who followed him, including relief from their pain and suffering for all eternity in God's paradise. Quoting profusely and enthusiastically from the Bible, the self-appointed prophet, usually a person with strong personal magnetism and great eloquence, quickly won the hearts and souls of the desperate.

The people of the Middle Ages, overwhelmed by very bleak conditions, lived in "a more or less constant state of apocalyptic expectation."[14] Much has been said about how these expectations exploded into generalized panic around the year 1000 C.E., the "terrors of the year 1000." However, a more careful analysis has shown that we just do not have enough evidence to determine what truly happened around 999—whether people's general state of mind was any worse than in other years. Many of the reports of Europe's coming to a halt as the fateful hour approached, which include tales of horrible famines and even cannibalism, are probably exaggerations concocted later for propagandistic purposes. One of the best-known sources from the time is the travel diary of the French nomadic monk Radulphus Glaber, a.k.a. Ralph the Bald, who paints a picture of absolute social chaos and despair, triggered by a series of cataclysmic events, ranging from a violent eruption of Mount Vesuvius in Italy to widespread pestilence and famine. Referring to the situation in Italy, Glaber writes,

> In those days also, in many regions, the horrible famine compelled men to make their food not only of unclean beasts and creeping things, but even of men's, women's, and children's flesh, without regard even of kindred; for so fierce waxed this hunger that grown-up sons devoured their mothers, and mothers, forgetting their maternal love, ate their babes.[15]

Glaber's account was not the only one painting the terrors of the year 1000. Centuries later, the preacher of the cosmic doom we met earlier in this chapter, Increase Mather, wrote in his *Cometography* of

> an horrendous comet seen appearing in the month of December, followed by a most terrible earthquake. And great wars between the emperor of Constantinople and the Bulgarians. . . . Also a grievous famine attended this comet, and a prodigious drought, so as that both Men and Cattle died of thirst.[16]

The recent *Cometography* by the American astronomer Gary W. Kronk makes no mention of this "horrendous comet."[17]

The year 1000, if not a year of singular despair in European history, was certainly the beginning of a new era. Maybe the changes that followed the turn of the millennium were the product of a collective sigh of relief from the postponement of the end. Pope Urban II quickly seized the opportunity, summoning the faithful to jointly take up arms in what came to be known as the First Crusade—that momentous attempt to regroup Christianity, liberate the Eastern Empire from the Muslims, and stop the wars ravaging most of feudal Europe. His famous 1095 appeal at Clermont sent waves across Europe. Noblemen and commoners alike vowed to serve Christendom. The crusade was transformed into a battle for spiritual salvation and, of course, a little material and political gain on the side. It galvanized the millenarian expectations of the poor, who left in droves to march to Jerusalem in search of holiness. Bearing the cross sewn on their crude outfits, the army of Christ, blinded by its lofty goals and charismatic prophet leaders, "liberated" Jerusalem by massacring every man, woman, and child who refused immediate conversion. That included not only Muslims but also thousands of Jews unlucky enough to be on the path of the "liberators." One account includes a horrifying scene of horses wading "in blood up to their knees, nay up to the bridle."[18] The killings were all justified by faith, as is often the case in "holy wars," believed to be a necessary part of the mission in bringing about everlasting peace. The religion that preached brotherly love was turned into a demonic device of destruction. And what is worse, in the minds of the "holy" murderers, the end perfectly justified the means: the blood of the sinner is the sanctified wine of the just.

The Masque of the Red Death

The church grew stronger. Gothic cathedrals were erected in France and Germany, the crusaders kept fighting for Christendom and salvation in the East and the Iberian peninsula, while the clergy exerted its dominance over the souls of people and kings. As commerce expanded, many peasants flocked to the growing cities in search of jobs and protection. Europe was caught completely unprepared for the challenges that lay ahead: growing commerce and lack of urban sanitation also meant the rapid spread of disease and poverty. Confined to small villages or farming communities in the early Middle Ages, the pestilence assumed epic proportions in the centuries to come.

In 1186, an astrological forecast brought normal activity in much of Europe to a halt; a major planetary conjunction was to take place in the constellation of Libra. Since Libra is an "air" sign, the conjunction was interpreted as a sign of, among other evils, a cataclysmic windstorm like none ever seen before. Ensuing earthquakes were to complete the destruction. German chronicles relate that people dug underground caves where they intended to hide during the storms, while several special services were held in churches to appease the panic-stricken population. It was predicted that in sandy regions, such as Egypt and Ethiopia, entire cities would disappear. An astrologer named Corumphira wrote,

The conjunction about to take place, whatever others may say, signifies to me, if God so wills, the mutation of kingdoms, the superiority of the Franks, the destruction of the Saracenic race, with the superior blessedness of the religion of Christ, and its special exaltation, together with longer life to those who shall be born hereafter.[19]

The forecast translates well the overall emotional climate in Europe, where people of all levels of society were more than willing to embrace any prophecy promising radical change. Against this background of fervent but unfocused expectation, the Calabrian monk Joachim of Fiore (ca. 1135–1202) brought about a renewed convergence of apocalyptic prophecy with his prognostications of an impending new age, complete with an estimate of dates. A skilled numerologist, Joachim used the Bible as a predictive instrument, which divided history into three eras, imprints in time of the Holy Trinity: the Old Testament, or the age of the law, was the age of the Father, when God's heavy hand ruled over the world's affairs; the New Testament, or the age of grace, was the age of the Son, marked by the establishment of the Catholic Church; the new age to come was the age of the Holy Ghost, an age of renewed spirituality, when Christianity's loftiest ideals would finally be realized and humankind would live in a permanent state of ecstatic religious contemplation and purity. One hundred years after Joachim's death, the great Italian poet Dante Alighieri placed him in paradise.

Joachim's intentions were far from revolutionary or subversive; he had the blessings of three popes and was widely admired. And yet, his ideas generated a series of movements that threatened the stability of the church. His complex calculations were based on Saint Matthew, who said forty-two generations passed between Abraham and Christ. Assuming that the same period of forty-two generations separated the second and third eras, and estimating each generation to last thirty years, Joachim placed the beginning of the third era somewhere between 1200 and 1260. Time was running out.

History, through the eyes of prophecy, becomes the foreseer of things to come, a reversal of its normal role of looking at past events. The people of the late Middle Ages saw life as the unfolding of the great apocalyptic drama, each individual identifying with one of several characters as he or she saw fit: saints and martyrs, angels and devils, messiahs and Antichrists. It was as if reality turned into a huge wave, which inexorably dragged existence bubbling and foaming along with it, until the inevitable crash at the beach.

This is when the flagellants appeared. Surging through Italy around 1260 and moving north to Germany, these bands of penitent people took the saying "no pain no gain" to its extreme: to them, self-inflicted physical pain was a purging of the soul, a declaration of religious fervor and martyrdom, a one-way ticket to salvation on judgment day (see figure 5 in insert). The flagellants were the chorus line of the apocalyptic drama, enacting a terrifying choreography of despair. Imagine a parade of hundreds, sometimes thousands of people wearing white hooded robes with red crosses sewn on the front and back, carrying banners and candles while singing religious hymns of repentance, marching from village to village, urging passersby to join in if they wanted to be saved from eternal doom. After arriving in a village, the flagellants would circle its main church or square and start their ritual of self-flogging, using leather whips indented with iron spikes, crying and wailing deliriously for eternal salvation while blood squirted from their mangled bodies.

The peak of the flagellant movement coincided with, or was a response to, one of the most horrendous episodes in the history of humankind, the Black Death. This epidemic of bubonic fever, which apparently originated in Central Asia and first appeared in the crowded port cities of southern Italy in 1347, spread like brushfire through the countryside and across northern Europe, killing about a third of its population in under two years. It is estimated that over twenty-five million people died in Europe and another fifty million in Asia, more than during the two world wars of the twentieth century.

The signs that such devastation was possible were plentiful. In 1340, a plague killed about fifteen thousand people in Florence. Plagues and pestilence were reappearing in overpopulated cities with growing fury; a total lack of sanitation, practically nonexistent habits of personal hygiene, and a tendency to accept disease as a chastising of sinners did not help the situation. The Black Death was regarded by many as a second flood, perfectly consistent with the approaching end. To lend further credence to its role in the apocalyptic drama, the ever-watchful perceived an abundance of cosmic signs. The chronicler Giovanni Villani, in his *Florentine Stories* (1348), blamed the 1340 Florentine plague on a comet, which appeared at the end of March in the constellation of Virgo. In the *Chronicle of Jean de Venette* (1368) of Gaul, the same comet, with "tail and rays extended toward the east and north," was "thought to be a presage of wars and tribulation to come in the kingdom."[20] The Hundred Years' War (1337–1453) had just started its periodic waves of devastation. Villani recorded another comet in 1347, which was to bring on the death of kings. In the *Decameron*, Giovanni Boccaccio described the arrival of the Black Death in Florence, uncertain whether the epidemic appeared "through the operation of the heavenly bodies or because of our own inequities which the just wrath of God sought to correct." To most people, the two were related. French texts tell of a star that hovered over Paris during August, while the German *Nuremberg Chronicles* (1493) mention a "hairy star" that remained visible for two months (see figure 6 in insert). That these celestial objects could also be seen in other times did not detract from their collective evocation of impending doom. In the meantime, corpses accumulated in fetid piles faster than they could be buried, while widespread famine compounded the devastation caused by the merciless killer. As the panic increased, so did the crowds of parading flagellants, whose infected open wounds promoted the spread of the bacillus. Cries of "Bring out your dead! Bring out your dead," followed by the sad tolling of bells, echoed in every town and village. While people died like flies in the streets, members of the high clergy and the aristocracy barricaded themselves behind the fortified walls of their castles, waiting for the wave of death to pass by.

In his allegorical tale inspired by the Black Death, "The Masque of the Red Death," Edgar Allan Poe (1809–1849) writes of the tragic end of Prince Prospero and his aristocratic guests, who arrogantly believed their lives were protected from the epidemic by their higher social status. They danced and dined within the walls of Prospero's castle while the peasants died outside, begging the indifferent prince for protection. Death makes an entry into the castle during a masquerade: "His vesture dabbed in *blood*—and his broad brow, with all the features of the face, was besprinkled with the scarlet horror." The dancing and laughter stop and dread spreads through the castle, as one by one Prospero's guests fall to the ground, the once bustling rooms turned into ghostly chambers:

And now was acknowledged the presence of the Red Death. He had come like a thief in the night. And one by one dropped the revellers in the blood-bedewed halls of their revel, and died each in the despairing posture of his fall. And the life of the ebony clock went out with that of the last of the gay. And the flames of the tripods expired. And Darkness and Decay and the Red Death held illimitable dominion over all.[21]

Re-creating Heaven and Hell

Faced with the horrors of the Black Death and the succession of devastating wars and famines, people understandably yearned for salvation, seeing their tragic lives within the framework of a grander apocalyptic plot. This transposition of biblical and mythic characters into real life found dramatic expression in the artistic output of the times: apocalypticism was very pervasive, an integral part of the lives of everyone, not just the clergy and the fanatic.

Great works of literature from the late Middle Ages, such as Dante's *The Divine Comedy* (1321) and Chaucer's *The Canterbury Tales* (ca. 1370), and from the Renaissance, such as Spenser's *The Faerie Queene* (ca. 1590) and Milton's *Paradise Lost* (1667), abound with apocalyptic references, mixing the fantastic with the real, bringing characters from politics and clergy and often the poets themselves into the narrative. Medieval and Renaissance painters did the same: apocalyptic iconography became one of the main sources of inspiration to many artists of the time, from the great Italian masters Giotto di Bondone, Luca Signorelli, and Michelangelo to the Germans Matthias Gerung and Albrecht Dürer. In their paintings, frescoes, and engravings, we find both aspects of the apocalyptic narrative, the tragic and the hopeful, often combined. Images of the last judgment, or the fall of the Antichrist, are so abundant that I can mention only a few representative examples. Signorelli's amazing frescoes in the chapel of San Brizio in Orvieto's cathedral, depicting the Antichrist preaching, display several stages of the apocalyptic drama. We see cosmic signs of the end of the world, the resurrection of the dead, the consignment of sinners to hell and the ascent of the just to heaven. In the fresco *Deeds and Sins of the Antichrist,* a group of Dominicans argue behind Antichrist, who is preaching from a pedestal while the devil whispers in his ear. The group of listeners to his right includes Dante, Signorelli himself, his mentor Fra Angelico, and other important Florentines of the time (see figure 7 in insert). The cosmic and the contemporary are clearly fused, lending reality to the prophetic narrative. In the Vatican's Sistine Chapel, the *Last Judgment* by Michelangelo (finished ca. 1541), portrays the saved and the damned side by side, light and darkness balancing the two parts of the fresco in a profusion of images of punishment and bliss.

The Germans were not as subtle. Dürer's collection of engravings from 1498, *The Apocalypse,* is full of the most horrific imagery, largely built upon dramatic cosmic events. In *The Opening of the Fifth and Sixth Seals*, a shower of stars falls over the heads of terrified sinners below, while a darkened Sun and a bloodied, angry-looking Moon decorate the lower skies. High above the chaos, we see the celestial throne surrounded by martyrs and angels, on their faces an expression of beatitude and peace. High members of the Catholic clergy are depicted among the sinners below (see figure 8 in insert). This collection of engravings predates the start of the Reformation by only nineteen years. It was in 1517 that Martin Luther nailed the ninety-five theses on the door of the church in Wittenberg, condemning, among other things, the power of priests to pardon sins through confession and absolution. The Catholic Church and its leaders were seen by many as completely corrupt and decadent; the reformists condemned popes and bishops with illegitimate children, their wealth and armies blatantly at odds with a religion that preached humility and peace. There was a proliferation of horrifying images of popes as Antichrists, popes crucifying Christ, devils excreting popes, popes as devils, and so on.

In the literature of the late Middle Ages, we find similar messages, albeit usually presented in a somewhat more discreet format. Dante's *Divine Comedy* and Chaucer's *Canterbury Tales* are structured as pilgrimages. Although Chaucer's often comic masterpiece is much lighter in content than Dante's journey through the afterlife, it nevertheless concludes with the arrival at Canterbury—which represents "Heavenly Jerusalem"—where resides the "holy blissful martyr, quick to give his help to them when they were sick." Chaucer treats the members of the clergy among the pilgrims with obvious sarcasm: of the monk, who had a "special license from the Pope," he writes,

> Sweetly he heard his penitents at shrift
> With pleasant absolution, for a gift.
> He was an easy man in penance-giving
> Where he could hope to make a decent living.[22]

The cynicism directed at the church gets worse as the thirty-one pilgrims tell their stories. Dante's rhetoric is more subtle and more somber, although no less efficient. There are several images borrowed straight from Revelation, as well as condemnations of popes as Antichrists. The poet often assumes the role of the prophet, denouncing the ills of society and clergy as he climbs from the searing heat of hell to the blinding light of paradise. Dante's personal life and tribulations are painfully reflected in his poem. If in real life he was exiled from his beloved Florence because of political intrigue, in the poem he faces another exile, that of the afterlife. If in real life his love for a woman called Beatrice was condemned by social convention, in the poem it is Beatrice who guides him through paradise. Dante kept his love secret and never actually touched the woman he so fervently desired, who, much to his despair, died very young. The poem joins his unconsummated love for Beatrice, shared by all unquenched lovers, with the unconsummated love for God, with which every Christian wrestles. In so doing, Dante masterfully blends the two main concerns of his times, true love in life and salvation in death.

A profusion of art has been created around the theme of loss, in particular the death of a loved one. The helplessness of losing someone dear as a result of uncontrollable factors, the despair of knowing we will never touch or be

touched by this person again or, as the twentieth-century Italian writer Piran-
dello said of his mother, of knowing that person no longer will be able to think
of you is, sometimes, too much to bear. Death sharpens all the corners we des-
perately try to round out by our humanity. What can we do but create our own
poetic justice, crying out our despair in poems and paintings, symphonies and
plays. Sometimes the stories end in tragedy, as in Shakespeare's *Romeo and
Juliet* (ca. 1595), which so blatantly explores the inevitability of fate. This theme
reemerges in his Sonnet 14, full of cosmic symbolism:

> Not from the stars do I my judgement pluck,
> And yet methinks I have astronomy;
> But not to tell of good or evil luck,
> Of plagues, of dearths, or seasons' quality;
> Nor can I fortune to brief minutes tell,
> Pointing to each his thunder, rain and wind,
> Or say with princes if it shall go well
> By oft predict that I in heaven find.
> But from thine eyes my knowledge I derive,
> And, constant stars, in them I read such art
> As truth and beauty shall together thrive,
> If from thyself to store thou wouldst convert.
>   Or else of thee this I prognosticate;
>   Thy end is truth's and beauty's doom and date.[23]

The sonnet yields a double blow on our vain search for truth and beauty. While
deriding beliefs in the powers of astrology, or in his ability as an astrologer,
Shakespeare laments his despair, knowing that the truth and beauty he sees in
his lover's eyes—the poet's astronomy—are ephemeral, as they cannot be as
eternal as stars. He finds solace in his verse: only through creativity can we
defeat death, as the closing lines of Sonnet 19 defiantly proclaim:

> Yet, do thy worst, old Time! Despite thy Wrong,
> My love shall in my verse ever live young.

Another great classic of late English Renaissance literature, John Milton's *Paradise Lost,* finished in 1667, delves deeply into immortality and salvation, resort-
ing often to cosmic symbolism. Milton was the son of a Protestant father who
had in turn been disinherited by his own fervently Catholic father. As the ten-
sions of the Reformation ran deep in the family, Milton sided with his father,

becoming an ardent critic of the Catholic Church, constantly condemning its corruption and equating the pope with the Antichrist. To Milton, and to the reformists, the fact that the Catholic Church gave its followers the choice of absolution through confession and repentance was unacceptable. This structure gave enormous power to the clergy and was at the heart of its corruption; as Chaucer's mockery of the pilgrim Monk made clear, priests were more than happy to exchange absolution and eternal salvation for a gift or two. The thematic structure of Milton's masterpiece poem, *Paradise Lost*, revolves around the first choice ever made in (Judeo-Christian) history, the choice by Adam to taste the forbidden fruit from the Tree of Knowledge, thereby contradicting God's orders, and the chain of events this choice unleashed. Adam's choice was clearly a bad one, its punishment being human mortality. As we read in the opening lines of *Paradise Lost*,

> Of mans First Disobedience, and the Fruit
> Of that Forbidden Tree, whose mortal tast[e]
> Brought Death into the World, and all our woe,
> With loss of Eden, till one greater Man
> Restore us, and regain the blissful Seat. . . .

We are all paying for Adam and Eve's transgression, at least until we reembrace eternity at the end of time. Both the sinner and the virtuous regain their lost immortality, the former in the torments of hell and the latter in paradise. What happens after death depends on the choices we make in life.

Milton's message came charged with cosmic signs. In his travels after finishing his studies at Cambridge University, he met Galileo, "grown old, a prisoner to the Inquisition for thinking Astronomy otherwise than the Franciscan and Dominican licensers thought."[24] His hatred of the Catholic Church was reinforced by seeing this great scientist fall prey to Franciscans and Dominicans, the two groups that, four centuries earlier had been considered the messengers of the new age predicted by Joachim of Fiore. The true story of Galileo's conviction by the Inquisition is much subtler, as I explained in my book *The Dancing Universe*. However, Milton's visit to Galileo was an expression of his interest in the changes occurring in the science of the heavens. In *Paradise Lost*, Milton blends the science of the cosmos with the cosmic religious fantasy of Revelation, invoking sulfurous storms and fiery cataracts descending from a "Firmament of Hell," red lightning, and darkness, side by side with references to the Sun's relative position with respect to several constellations. In the closing lines, he compares God's sword of light to a comet:

The brandisht Sword of God before them blaz'd
Fierce as a Comet; which with torrid heat,
And vapour as the Lybian air dust,
Began to parch that temperate Clime. (book 12, lines 633–36)

From Dante to Milton, from Signorelli to Dürer, the artistic output of the late Middle Ages and the Renaissance reflects the obsessive fascination that apocalyptic themes, and their link to celestial phenomena, exerted on European society. These works created a databank of cosmic apocalyptic images, a common set of graphic representations and literary symbols of eschatological themes that have been an integral part of the collective imagination of humankind ever since. As we will soon see, they greatly influenced early astronomical thought, because the central quest of the natural philosopher was to unravel God's messages written in the skies. And, perhaps surprisingly, they continue to influence astronomical thought today, even if dressed in the quantitative language of modern science.

Apocalypse Now!

We still, and apparently at an accelerated rate, speak of eternal salvation and damnation, of a fiery hell below and a paradise above, of cataclysmic events as punishments from God, of unpredictable cosmic phenomena as the heralds of doom or, as in a profusion of enormously popular recent movies, of comets and asteroids as the instruments of doom themselves. A 1983 Gallup poll disclosed that 62 percent of Americans had "no doubt" that Christ will return to Earth sometime in the future. In 1992, about 53 percent believed that the return was imminent and that it would be followed by the cataclysmic events connected with the destruction of evil as prophesied in the Bible. In Canada, a 1993 poll revealed that, no less than 30 percent of the respondents believed it will happen within the next hundred years. Our collective eschatological imagination is as active as ever, even if its symbolism has been greatly influenced by science. We still fear what we cannot understand and control; thus, as scientific progress shifts our ignorance about nature to different areas, we also shift our fears. In order to illustrate how this shifting operates, we will briefly examine the tragic history of some apocalyptic sects and their beliefs, past and present.

During the early years of the Reformation, widespread feelings of despair and distrust spawned an abundance of apocalyptic sects. People were desperate for salvation at any cost. The signs were there for all to see, from the accusations of popes and Catholic prelates as being Antichrists, to famines, plagues, wars, invasions, and, of course, comets and other unusual cosmic phenomena. In 1524, astrological forecasting of a flood to be "caused" by a conjunction of all planets in Pisces was taken so seriously that a magistrate from the parliament of Toulouse ordered an ark to be built on top of a nearby hill. The very rupturing at the seams of the Catholic Church by the Protestant "rebellion" was seen by many as a sign of the approaching end. The Protestant reformers Martin Luther and Philipp Melanchthon (1497–1560) were convinced the end would happen before the 1600s. In 1522, Luther preached a sermon on the Second Coming, declaring that the impending doom was written in the skies:

> We see the Sun to be darkened and the Moon, the Stars to Fall, men to be distressed, the winds and waters to make a noise, and whatever else is foretold of the Lord, all of them come to pass as it were together. Besides, we have seen not a few Comets, having the form of the Cross, imprinted from heaven. . . . How many other Signs also, and unusual impressions, have we seen in the Heavens, in the Sun, Moon, Stars, Rain-bows and strange Apparitions, in these last four years? Let us at least acknowledge these to be Signs, and Signs of some great and notable change.[25]

As countless prophets appeared everywhere, the Protestant upsurge inspired more dissenting sects than even the reformist leaders could digest. Among these, the most driven were the Anabaptists, who repudiated both the Catholics and the Lutherans, commanding their followers to rebaptize. Occasionally, one would be seen running naked through the streets of Amsterdam and crying, "Woe, woe, woe the wrath of God!" The 1525 peasant revolts in Germany were believed to be the battle of the saints during the last days—until they were completely crushed and their Anabaptist leaders executed. The return of Halley's comet in 1531 compounded the grim expectations. In 1534, the German town of Münster was shaken by a storm of Anabaptist militancy. Within a week of preaching in the streets, more than fourteen hundred people were rebaptized. A young Dutch preacher named Jan Bockelson, known as John of Leiden, caused a frenzy of apocalyptic fanaticism whenever he spoke. Women, including many nuns who quit their convents to "convert" to the new faith, would undergo fits of hysteria, throwing themselves on the ground, screaming, kicking, and foaming at the mouth. Eventually the Anabaptists invaded the town hall, causing a mass exodus of the Lutheran majority. They invited all neighboring Anabaptists to join them in Münster, the "New Jerusalem," which alone would be saved from destruction at the last days.

Soon the Anabaptist government turned to terror; it confiscated property, banned all books except the Bible, and tortured and executed dissenters. John of Leiden instituted polygamy, and declared himself king of New Jerusalem, none other than the Messiah foretold by Old Testament prophets. Funded by heavy taxation, the new court quickly grew sumptuous, while the people of Münster experienced abject poverty and privation. "Not to worry," John would tell his disciples, "soon you too will be covered by the gold and silver of paradise." At that point, the town was already under siege by an army of mercenaries organized by Münster's former Catholic bishop. He managed to get help from many local rulers, and in January 1535 a complete blockade was instituted, slowly starving the inhabitants of New Jerusalem. In a few months, as the promised bread did not fall from heaven, all animals were eaten, from dogs and horses to rats and hedgehogs. Then people started to consume grass, moss, and the bodies of the dead. The king remained completely oblivious, until bodies began to litter the streets. Finally, he allowed those who wished to leave town to do so, although no better luck awaited them outside the fortified walls: they all met gruesome deaths at the hands of the mercenaries, who did not spare women, children, or the elders. In June 1535, the town was finally taken over, most of its battered population killed in a massacre that lasted several days. John was chained and paraded by the bishop on a leash, ordered to act like a bear for the amusement of mocking crowds. He was brought back to Münster early in 1536 and, with two other Anabaptist leaders, was publicly tortured to death with red-hot irons. It is said that the king did not utter a word or make a movement until his death. The three bodies were then left to rot in iron cages suspended from a tower of Saint Lambert's Church, a very gruesome sight still there to be seen today.

This horrendous episode sounds at first like a historical curiosity, a singular tale of apocalypticism run crazy. But it isn't. The last few centuries have seen many similar instances where a millenarian movement, initially with perfectly pacific intent, walled itself off from the outside world, its members psychologically (and sometimes physically) enslaved by a charismatic leader bent on taking his self-appointed prophetic mission to the end. The results have almost always been tragic. From the point of view of the "walled-in" sectarians, the unavoidable clash with the "outside world" is the battle of Armageddon, their martyrdom a passport to heaven. From the point of view of the outside world, the sect represents a threat to society, fueled by its anarchic mistrust of the government and backed up by the illegal possession of firearms and weaponry.

Skipping a few centuries of history, I will mention just a few recent examples of apocalyptic sects that bring back to life elements of the Münster insurrection of 1534–35. In 1993, the "walled-in" tradition of Münster was tragically revived in the conflict between the Branch Davidians of Waco, Texas, and the forces of the U.S. government. Their leader, David Koresh, was to be the opener of the seven seals of Revelation 6 and 7, to explain their meaning, and to lead his followers to heaven, transported in God's flying saucers, the modern-day version of God's chariots of fire of Psalm 68:17. After a fifty-one-day siege, which led to the death of several federal agents, the whole structure housing the sect members went up in flames, killing 73 men, women, and children, together with Koresh. Fifteen years before the Waco tragedy, 913 followers of James Warren Jones, a charismatic preacher who fled the United States to found his own walled-in New Jerusalem, a.k.a. Jonestown, in Guyana, committed collective suicide by drinking cyanide-laced Kool-Aid. Nuclear holocaust was one of his favorite themes.*

The mass suicides of the Branch Davidians and of the inhabitants of Jonestown tragically illustrate the deadly power of apocalyptic lore. The reader should not be quick to dismiss these acts as the follies of maddened religious radicals, or the product of ignorant superstition. Well-educated people, perfectly in tune with their times, often succumb to apocalyptic paroxysms. The visit of Halley's comet in 1910 was greeted with a mix of anxiety and humorous enthusiasm. *Harper's Weekly* published a drawing with the caption "Waiting for the End of the World," echoed by many counterparts in Germany. Other cosmic signs corroborated the grim expectations of the increasingly terrified population: unusual weather, more comets, meteor showers, sunspots, awesome displays of aurora borealis, and even the death of a king, Edward VII of England. Although there were no predictions of a collision with Earth, astronomers noted that the comet's tail, rich in cyanogen gas, would pass through Earth. Terror ensued when, among others, the French astronomer Camille Flammarion pointed out that the mixture of the comet's cyanogen with the hydrogen in our atmosphere would produce the deadly prussic acid. Gas masks and "comet pills" were quickly sold out, churches filled up with people desperate to purge their sins before "impact," doors were sealed up, and panic skyrocketed. On

---

*As I was putting the final touches on this chapter, another apocalyptic sect met with a tragic end, this time in Uganda, Africa. The number of bodies consumed by the fire that burned their place of worship keeps climbing, and has surpassed even the Jonestown count, with over 924 dead. Apparently, many of the deaths were not voluntary, but violently inflicted upon the victims, probably those who tried to flee from the burning church.

May 18, 1910, the *New York Times* reported that "terror occasioned by the near approach of Halley's comet has seized hold of a large part of the population of Chicago." New York was no exception; many of its citizens could be seen praying on their knees in parks and streets, while "several religious processions took place in different parts of the city."[26] In France, after a nightlong vigil charged with anxiety, people embraced and danced in the streets after the authorities declared the "danger" to be over. Of course, the trace amounts of cyanogen gas in the comet's very diffuse tail were never a true danger but a clear illustration of apocalyptic fear inspired by irresponsible apocalyptic science.

In contemporary society, apocalyptic jargon incorporates science's latest findings and cutting-edge technology. If angels are not coming from heaven, then flying saucers are, populated by benevolent, angel-like aliens, such as the ones featured in Steven Spielberg's *Close Encounters of the Third Kind* or his extraterrestrial emissary of love in *E.T.* A comparison can be made between the alien in *E.T.* and Jesus himself. After all, the alien could perform countless miracles, including bringing dead things back to life, communicate telepathically, fly and make others fly, resurrect with a big red beating heart showing through his chest (as in many depictions of Christ), and, finally, ascend back to heaven in one of God's modern "chariots of fire." What were once supernatural powers attributed to gods are now "natural" powers attributed to aliens, expressing a deep connection between religious symbolism and our age-old expectations inspired by scientific ideas. Hence Spielberg's genius, of tapping into our collective expectations of redemption, using the modern language of science and its fictions.

Extraterrestrial angels were also believed to be the saviors of another apocalyptic sect, known as Heaven's Gate. Here too the story ended in tragedy (at least from the point of view of the "outside world"), when thirty-nine members were found dead on a ranch near San Diego, California. The precursor event for the collective suicide was the passage of the comet Hale-Bopp during the winter and spring of 1997. The sect's leader, Marshall Applewhite, or "Do," believed that his soul and those of his followers were to be hoisted from their earthly existence by a flying saucer coming from the "Level above Human," which was hiding behind the comet. Their destiny was to share a bodiless eternity in the "Kingdom of Heaven," among the stars and planets. Unlike those of most other apocalyptic sects, the members of Heaven's Gate were not left behind by modern technology; rather, they were deeply unhappy with what technology and modern life had to offer. The apocalyptic thread they shared with other sects could be found in their deep criticism of what they saw as a corrupt world, whose religious institutions had been seized by Lucifer, or "Lucy," as Do liked to call Satan.

Their expressed dislike for their bodies matches well their connection with the Internet, a high-tech version of a disembodiment ritual: only your message—your spirit—travels across the web, connecting with other "spirits" in far-removed links. When an amateur astronomer "saw" an unidentified point of light (the long-awaited redeeming UFO, which was soon afterward revealed to be the star SAO 141894) trailing the comet, Do knew the time had come. On March 23, during the comet's closest approach with Earth, the first wave of fifteen suicides took place, soon followed by the rest. Apparently, they all died willingly and happy, eager to embrace their new form of existence.

Thus Spoke the Science Apologist

It seems somewhat paradoxical that at the start of a new millennium, during what we proudly refer to as the scientific age, many people still look back to the prophecies of Revelation with such fearsome awe. There is growing cynicism toward science, a sense of betrayal, of promises unrealized. After all, was not science supposed to be the new redeemer, the accumulated knowledge of the world, our shining sword to ward off the threats of unpredictable nature? We get cures for myriad diseases only to discover new, incurable ones; we create new technologies that supposedly make life easier and more pleasant, only to spend more hours than ever at work. Even worse, technology advances so fast that it is virtually impossible for most of us to keep up, and a vast "technological underclass" is emerging, reminiscent of the socially displaced rural migrants in the medieval cities. We can send a man to the Moon (or could, when it was politically relevant) but cannot feed most of the world's population. We consume the natural resources of our planet with an appetite worthy of one of Daniel's apocalyptic beasts, feeding our endless greed for material goods without looking back at the devastation we often leave behind. And all this thanks to "science"!

So goes the credo of the discontent. Now I must wear the robes of the Science Apologist and refute all the above accusations.

"First and foremost, science does not promise redemption. Science is a human invention preoccupied with understanding the workings of nature. It is a body of knowledge about the universe and its many inhabitants, living and nonliving, accumulated through a process of constant testing and refinement known as the scientific method. What the practice and study of science does provide is a path back to nature, a way of reintegrating ourselves with the world around us. In so doing, it teaches us that the essence of nature—from the inanimate to the animate—is change and transformation, that life and death are intertwined in a cosmic chain of being. It was the 'death' of a nearby star that triggered the formation of our Sun, where life became possible in at least one member of its court of planets and moons. If there was life near that original dying star, it was destroyed with it, the same way life here will be destroyed when our Sun burns out. This dance of creation and destruction is constantly happening throughout the universe, linking our histories, our lives and deaths, to a larger cosmic chain of transformation. As such, every link is important, from what we create and destroy in life to what we leave behind. Science may not offer eternal salvation, but it offers the possibility of a life free from the spiritual slavery caused by an irrational fear of the unknown. It offers people the choice of self-empowerment, which may contribute to their spiritual freedom. In transforming mystery into challenge, science adds a new dimension to life. And a new dimension opens more paths toward self-fulfillment." Thus spoke the Science Apologist.

"Second, science does not determine what is to be done with its accumulated knowledge: we do. And this decision often falls into the hands of politicians, chosen by society, at least in a democracy. The blame for the darker uses of science must be shared by all of us: Are we to blame the inventor of gunpowder for all the deaths from gunshots and explosives? Or the inventor of the microscope for the development of biological warfare? We, the scientists, have the duty to make clear to the public what we do in our labs, and what consequences, good or bad, our inventions may have for society at large. But there is no such thing as 'the scientists' as a group that shares a set of morals or views, or the blame for the uses and abuses of science. There is, I would like to believe, a common set of goals, to better understand the world and our place in it and, yes, to improve our living conditions and health. It has been argued that defense brings peace, that the accumulation of an arsenal of destruction wards off further armed conflicts, at least large-scale ones. We created a war without winners. Thus, many in the defense and weapons industries see themselves as wardens of the peace and not creators of weapons of destruction. Personally, I see the need to collect weapons to guarantee the peace as a sad confirmation of our collective stupidity." Thus spoke the Science Apologist.

"Finally, science has not betrayed our expectations. Think of a world without antibiotics, computers, televisions, airplanes, and cars—a world in which we are all back in the forests and fields where we came from, living with no technological comfort. How many of us would be ready or willing to do it? Can you see yourself living in some cave or primitive hut, hunting for food, physically fighting constantly for survival? There is much hypocrisy in the criticism of science and of what it has done to us and to the planet. We did it all ourselves, through our choices, creativity, and greed. It is not by slowing down scientific research or its teaching, through legislation or censorship, that we will change the inequities of a technological society; that is surely a one-way ticket back to the Middle Ages. What is needed is universal access to the new technologies, an 'internetization' of society at large, coupled with a widespread effort to popularize science, its creations, and its consequences. Only a society well versed in scientific issues will be able to dictate its own destiny, from the preservation of the natural environment to the moral choices of genetic research.

"What happened to telephones and televisions will also happen to the newest technologies; they will become (almost) universally available. But there will always be a lag time, and this delay ostracizes much of the lower-income population or those without access to the latest innovations. From this 'access gap' is born the modern version of a technological underclass, deeply mistrustful of those who control information and its production and dissemination, even though they may use it, as in the case of the Heaven's Gate sect. Anxieties soar, and the idea that conspiratorial groups are plotting to take over the world becomes plausible: movies such as *Wag the Dog* describe the fabrication of realities by the news media; hugely popular sci-fi television shows such as *X-Files* have plots based on a secret conspiracy involving a partnership between sectors of the U.S. government and aliens. Those fears are contrasted with the images of the 'successful people,' the beautiful stars, the rich and famous, the ones with access to all the new techno-toys you can possibly find. As a result, many people feel used and useless, mere spectators in a game they can never play. This situation, within limits, can be compared to what was happening in Europe in the thirteenth century, when a large urban underclass was developing. Now and then, a charismatic leader appears, promising salvation and redemption, a new life for those that follow him or her. Now and then, extremist religious movements are born, often blending Christian eschatology with technological and paramilitary elements, where the members of the group see themselves as the agents of change, the key players in the great apocalyptic drama. The 'access gap' widens into an abyss, anxiety becomes anger, and a hunger for justice blinds any vestige of social morals; in their minds, the final crusade is starting, and it must be fought to the end.

"Science, like much else, is completely helpless against this form of religious extremism. There will always be people who find no other path to spiritual salvation but that offered by an all-or-nothing kind of logic. Nevertheless, I believe there is hope. Science and religion should not be pitted against each other, but seen as twin paths to a better life. Their complementarity springs from their common source, our fascination with questions beyond our control and understanding. They both express our awe with things that are bigger (and smaller) than ourselves, attempting to expand our vision of the world within and without. Their methods are certainly different, as are their immediate goals, but most people need both. It would be a mistake to think that society could advance with a purely analytical or with a purely faith-based approach to existence." Thus spoke the Science Apologist.

PART II

Cosmic Collisions

CHAPTER 3

Making Worlds

All coming into being is mixture, all perishing dissolution.

— ANAXAGORAS OF CLAZOMENAE

(CA. 500–428 B.C.E.)

There was great commotion in Athens that summer morning. The year was 354 B.C.E., and the thirty-year-old philosopher Aristotle had told his fellow members of the Academy that a great debate was being organized in Delphi, home of the famed and mysterious oracle. "Why Delphi?" asked Juboxus, one of Aristotle's pupils, given to long contemplative bouts, inspired by nothing more than the whispering wind or the spirals of a seashell. Some historians consider him a follower of the philosopher Pythagoras, even though most of these number mystics had disappeared by then. "Ha!" exclaimed Aristotle without hiding his disdain. "Delphi has been chosen for being considered, at least by ignorant fools, a magical place, where the spirits of the old 'physicists' may speak to us through the oracle." "Fantastic!" exclaimed Juboxus, trembling with anticipation. "I would not get too excited with all this hocus-pocus," said Popicas, another of Aristotle's pupils, more focused and intense than Juboxus, but barely able to conceal his secret fascination with the unknown. In his spare time, he was keen on designing impossible cities that defied all principles of then known architecture. Popicas was a visionary urbanist, who openly defied most Greek architects' blind slavery to the straight line. "Sinuosity is the way to re-create the world in our cities, not boxy-looking things! Have you ever seen a boxy cloud? What is the line traced by the wingtip of a swallow? That should be the outline of our roofs!" Popicas was what we could call a theoretical archi-tect, an artist more than a builder. Little did he know that his style was to be finally implemented, some twenty-three hundred years after his death, by the Spanish architect Antoni Gaudí and others.

The trip to Delphi was charged with philosophical discussion. An experi-enced mentor, Aristotle would provoke arguments among his pupils by stating an impossible proposition. One of his favorites was "The Sun is a hot rock." Since students love to prove their mentor wrong—and this is an important step in learning, for it promotes self-confidence and dispels the myth of infallibility of authority—the response would be automatic. Everyone had his own way of disproving the master's statement, and a lively debate would ensue while Aris-totle patiently listened, nodding constantly, interjecting occasionally, and look-ing mildly impressed with their creativity. Like most mentors, he learned more from his pupils than he cared to admit. "But, Master," said Popicas, "you your-self told us that the Moon and all celestial objects above it, including the Sun, are made of a substance different from the four elements of our world—earth, water, wind and fire. Is it not called the fifth essence, or ether?" To this Juboxus replied, "The Sun, being an object of the heavens, is a perfect sphere. As such it cannot be made of the coarse material of earthly things. It hums its own melodies as it circles us in its celestial orb. Oh, how I wish I could hear it some-day!" "Hey, Juboxus, maybe if you just stayed in Delphi forever, you would end up hearing it," quipped Popicas. They accelerated their climb toward Delphi, nested on the southern slopes of fabled Mount Parnassus. Others were already waiting in the splendorous Temple of Apollo, where Talias, the oracle, was to conduct the séance-debate.

The meeting in Delphi was to center on one question of great interest to Greek philosophers, the plurality of worlds: Are there many worlds like our own in the cosmos, or are we alone? This question had been a hot topic of debate since the time of the first "physicists," a group of people who, starting with Thales of Miletus during the sixth century B.C.E., tried to understand the workings of nature within nature, without invoking the actions of gods or God. Aristotle had a great interest in their ideas, and a good fraction of his writing reviews and criticizes them. They are now known as the pre-Socratic philoso-phers, although some of them lived just after the birth of Socrates, around 470 B.C.E.

As Aristotle, Juboxus, Popicas, and other members of the Academy arrived, the place was already packed. People came from as far away as Syracuse in southern Italy and Ephesus in the west. The seating was arranged wedge-shaped around the altar, where the oracle was to preside over the ceremony.

Magnificent statues of the gods, painted to look eerily real, flanked the center stage, flowers and fruits adorning their marble feet. Excitement and the steaming of herbs made the air thick and moist, sparks ready to start flowing from head to head. Musicians played ancient melodies, said to have been composed by Pythagoras himself. Outside, vendors sold retsina and bread soaked with olive oil. Molten lead clouds swirled furiously overhead, while a strange mist emerged from the valley, as if escaping from an open wound on the earth's crust. Thanks to the amazing powers of Talias, the great oracle, two hundred years of knowledge were to coexist, the passage of time magically suspended. Philosophers from the Pythagorean school were to debate with those of the atomistic school and those with Plato and his followers, to the delight and instruction of the astonished audience. All voices flowed from the same source, the oracle who alone could tunnel in and out of the world of the spirits. Aristotle entrusted Juboxus with taking notes, which were to be discussed later in the Academy. As the beautiful Talias finally appeared on the altar, the audience fell silent, transfixed by her magnetic presence and majestic attire. Her hair was black as ebony, and her skin milky white. Over her white tunic, she wore a huge pendant encrusted with a shining ruby, supposedly the key to the other world. Rubbing his eyes constantly and pinching his somewhat plump belly, an awed Juboxus could hardly concentrate on his task. The text below, *The Stone of Aegospotami*, is what Juboxus managed to jot down on that memorable evening, complete with commentary added later.

The Stone of Aegospotami

Talias climbed on top of the famous cracked stones and took a long whiff of the yellow sulfurous vapors that seeped out. When her face emerged, she looked different, her eyes shining with an unearthly light. She was not Talias anymore, but claimed to be speaking for Anaximander, the successor of Thales, whom our master called the first philosopher. They lived about two hundred years ago [550 B.C.E.] in the town of Miletus, off the coast of Lydia [Turkey]. Anaximander said,

> Destruction and coming-to-be follow each other through the infinity of time. From the Boundless come into being all the heavens and the worlds within them. And the source of all existing things is where destruction, too, happens, *according to necessity; for they pay penalty and retribution to each other for their injustice according to the assessment of Time*.[1]

Talias's distorted voice faded, and her face changed back to normal. A growing buzz could be heard from the amazed audience. So this was Anaximander's teaching: that the cosmos is eternal and everything in it comes from this abstract quantity called Boundless and goes back to it in endless cycles. A nice dynamical picture of nature, if you ask me. This business of "penalty and retribution" fits neatly into his picture, for there must be a balance between creation and destruction. This is how injustice—the imbalance—is avoided as time passes. But people were interested mostly in Anaximander's use of the word "worlds" in the plural. Did he mean there were many coexisting worlds out there in the cosmos, being created and destroyed all the time, or that our world has been through many cycles of existence, flowing back into the Boundless just to reemerge again in the fullness of time? Although it is tempting to defend the first position, evidence points toward the second: that Anaximander thought only of our world's being created and destroyed in cycles. For him, the Sun, Moon, and stars were not other "worlds," but fire escaping from holes in cosmic wheels surrounding the Earth. It is curious, though, his idea of the Earth going through many existences. How did he envisage our world's destruction? Maybe he shared the image the great Plato later wrote in his *Timaeus*, when explaining how natural disasters of terrible consequences happen from time to time:

> *There is a story, that once upon a time Phaëton, the son of Helios, having yoked the steeds in his father's chariot, because he was not able to drive them, in the path of his father, burnt up all that was upon the Earth, and was himself destroyed by a thunderbolt. Now this has the form of a myth, but really signifies a declination of the bodies moving in the heavens around the Earth, and a great conflagration of things upon the Earth which recurs after long intervals.*[2]

Another connection comes to mind here. I remember that Anaximander's disciple Anaximenes conjectured that the "nature of the heavenly bodies is fiery, and that they have among them certain earthly bodies that are carried round with them, being invisible." About a hundred years after Anaximenes, similar ideas were echoed in the teachings of Diogenes of Apollonia. He also believed that many such bodies revolved unseen in the sky. It sounded to me like a very plausible suggestion, and I brought it up to the audience. Maybe these invisible bodies are the ones that fall on Earth, causing the destruction Plato mentions? Master Aristotle was quick to explain that, for him, heavenly bodies stay in the heavens; that rocks that sometimes fall on Earth came from the ground and not from the sky, being first removed by some violent force, like a storm, hurricane, or volcano eruption, and then returning to where they belong. Some of us were not as convinced, especially given the size of the famous "stone of Aegospotami," which fell there some 113 years ago [467 B.C.E.]. In fact, this event inspired the ideas of Diogenes I mentioned above. But Master Aristotle reminded us that the fury of nature should not be underestimated. With that, Talias returned to the altar and took another whiff of the foul-smelling stuff. Her body trembled, possessed by invisible waves, and, as she spoke, she uttered the words of Anaxagoras of Clazomenae:

> Some say I predicted the fall of the stone of Aegospotami. Of course, I did no such thing, although it honors me that people think I know the ways of the heavens. But some things I do know, that *the Sun gives the Moon its brightness, and the stars, in their revolution pass beneath the Earth.* The world came into being as *air and ether were being separated off from the surrounding mass, which is infinite in number. The dense and the moist and the cold and the dark came together here, where the Earth now is, while the rare and the hot and the dry went outwards to the further part of the ether. As these things rotated thus [they] were separated off by the force and speed. Their speed is like the speed of nothing that now exist among men, but it is altogether many times as fast.*

I like the idea that the Sun gives light to the Moon; Master Aristotle disagrees, for he thinks that all heavenly bodies, Sun, Moon, stars, and planets, are made of ether and generate their own light. But my favorite was Anaxagoras's mechanism for the formation of Earth. A primordial rotation separated the solid matter, which falls to the center, from the hot stuff, which drifted outward. The rotation generated a "force" that facilitated the compression of matter seeds into larger solid bodies. As far as I know, this was the first time someone proposed such a mechanism, and that alone deserves a lot of credit. The rest of the audience was not so impressed. But then again, some ideas, even if correct, take time to flourish. I think this is one of them.

Talias, looking paler than she already did before the ceremony, once more climbed the steps toward the sacred stones. Whom else could she possibly be "receiving"? I never thought the life of an oracle was this hard! She emerged from the vapors with an odd smile on her face. We then knew, even before she spoke, that she was Democritus, the atomistic philosopher known as the "laughing one." This is what "he" said:

> Everything is in constant motion in the void. There are innumerable worlds, which differ in size. In some worlds there is no sun and moon,

in others they are larger than in our world, and in others more numerous. In some parts they are arising, in others failing. They are destroyed by collision one with another. There are some worlds devoid of living creatures or plants or any moisture.

There was the idea of the cataclysmic collisions of worlds once again! Looking over my shoulder, I saw Master Aristotle squirming on his seat. Imagine that, other worlds with living creatures and plants! Democritus or, rather, his "voice" then explained how these worlds were formed, also through the swirling of matter, somewhat like Anaxagoras, but in much more detail. He and Leucippus believed everything was made of indivisible atoms moving in the void, which, through the revolutions of a kind of vortex, would congregate, like to like. It was something like a rotating disk of matter, where earthy atoms would concentrate in the center forming Earth, while others would rotate around it, coagulate with their neighbors, and ignite to become the heavenly bodies, after their original moisture dried up. A lot of the details were missing, but this was the general picture, at least what I could capture of it. Talias collapsed to the ground, and was taken to her private chambers by other priestesses. It was time for us to drink retsina, argue some more, and dance to joyous music. As we walked outside, we could not help looking upward, the countless stars saluting us for our intellectual courage. Or maybe they were mocking our ignorance?

Evil Also Rises

We can only marvel at the pre-Socratic imagination. Although, from our modern perspective, some of their ideas were quite off the mark, others were strangely prescient and are still with us, in one form or another. Today, the key word to describe the natural world is "change," from its smallest components to the universe as a whole. Even the idea of equilibrium, which seems to imply the absence of change, is often related to a dynamic balance of input and output; think of living systems, from single-cell bacteria to whales, which continuously take and give back to their surrounding environment. Sometimes the timescale for these changes is so long, compared with human scales, that we are fooled into thinking of stability. But mountains change, and stars move in the heavens. Leucippus and Democritus called the basic constituents of matter atoms; we call them elementary particles, but the idea that matter is made up of fundamental constituents is essentially the same. And from the fictional pilgrimage to

Delphi we learned that some pre-Socratics envisaged a cosmos with many coexisting worlds, some of them invisible to us, some empty, and others populated by animals and plants. The atomists, possibly inspired by Anaxagoras and Diogenes, went so far as to suggest a dynamical mechanism for the creation and destruction of worlds: worlds are created from the continuous aggregation of atoms induced by the rotation of a proto-disk of mixed matter, while worlds are destroyed by collisions with other worlds. These processes are random and purposeless, occurring in the "fullness of time." In the words of the Roman poet Lucretius, "Never suppose the atoms had a plan."[3] These ideas, as we will see, are incredibly modern.

But Aristotle would have none of it. His system of the world was radically different from that of the atomists. Change was relegated to the sublunary sphere, the heavens and its bodies being unchangeable, made of ether, the fifth essence (see figure 9 in insert). This division of the cosmos into two realms forced Aristotle to devise a very ingenious, if not fanciful, theory of comets and shooting stars. Since they were clearly time-dependent events, they had to belong to the earthly realm, where the four elements, with their combinations and motions, promoted the change we observe. His solution was to consider these celestial events as belonging in the upper atmosphere, the "outermost part of the terrestrial world which falls below the circular motion [lunary sphere]."[4] To Aristotle, comets and shooting stars were meteorological phenomena, caused by exhalations "of the right consistency" that rose up and were ignited by the dryness of the upper skies, being subsequently carried around by the circular motion of the nearby lunary sphere. Rapid, intense fires that burned quickly were the cause of the "'shooting' of scattered 'stars.' "[5] A comet, by contrast, was produced when a "fiery principle," intense enough to sustain its burning for a long time and tame enough not to be so bright, met the right kind of exhalation rising from Earth. The shapes depended on the burning patterns of the exhalation: "If [the exhalation] diffused equally in every side the star is said to be fringed, if it stretches in one direction it is called bearded."[6] The bearded exhalation "explained" the tails of comets.

Aristotle even related the presence of fiery comets to weather patterns such as droughts and strong winds. Like many of his successors, he viewed comets as portents, which could be used in weather prognostication. He illustrated his argument with the meteorite of Aegospotami, which he supposed had been swept up by winds caused by the presence of a comet: "For instance, when the stone at Aegospotami fell out of the air—it had been carried up by a wind and fell down in daytime—then too a comet happened to have appeared in the west."[7] Aristotle's theory of comets and shooting stars was consistent with his dismissal of the atomists' case for the existence of other, Earth-like worlds. Since these worlds were necessarily made of the four basic elements, they just could not fit into Aristotle's scheme of concentric spheres, where every element moved naturally in the realm where it belonged; other earthly worlds would all coagulate into this one, just as a stone falls back to the ground, where it belongs. We see, then, a split of opinions concerning the existence of other material worlds into mainly two groups: Aristotle and his followers explaining comets, shooting stars, and other sudden celestial apparitions as meteorological phenomena, Earth being the only material world subject to change; and the atomists, defending the existence of other material worlds, coexisting with our own, which sometimes may become visible as comets or shooting stars, or even collide with us. For Aristotle, comets foretold weather-related cataclysms; for the atomists, entire worlds might be destroyed by colliding with other worlds, which circle—sometimes invisibly—in the heavens.

Aristotle's views on comets, like much else he wrote, was to greatly influence astronomical thought well into the Renaissance. One discordant voice worth mentioning for its great intellectual courage was that of Seneca (ca. 4 B.C.E.–65 C.E.), the Spanish-born Roman orator and Nero's tutor, whose *Natural Questions* argued for a cosmic origin of comets. He shared many of the opinions of a pre-Socratic known as Apollonius of Myndos (fourth century B.C.E.), who claimed,

> Many comets are planets . . . a celestial body on its own, like the Sun and the Moon. . . . A comet cuts through the upper regions of the universe and then finally becomes visible when it reaches the lowest point of its orbit. . . . Some are bloody, menacing—they carry before them the omen of bloodshed to come.[8]

Seneca agreed with Apollonius that comets were celestial objects, but not that they were planets. After all, he argued, we can see stars through a comet's tail but not through planets. He believed they were objects of their own, "among Nature's permanent creations," to be approached with the reverence of things divine: "God has not made all things for man."[9] With incredible foresight, he speculated that comets might return in their celestial orbits, albeit at "vast intervals," and predicted, "The time will come when diligent research over very long periods will bring to light things which now lie hidden. . . . There will come a time when our descendents will be amazed that we did not know things that are so plain to them."[10] Seneca's lucid defense of a rational approach to the study of natural phenomena is tempered by his belief that comets served a divine purpose, which people could use for forecasting the future, if only they understood the message: "The roll of fate is unfolded on a different principle, sending ahead everywhere indications of what is to come, some familiar to us, others unknown."[11] He believed there was a divine cosmic plan, to which comets belonged. If we had access to its meaning, we could read the future like an open book.

Even those thinkers who claimed to defend a rational approach to the world could not shake themselves loose from the allure of comet-linked prognostication. Seneca died five years before the destruction of the Second Temple in Jerusalem, when Christianity started to spread its seeds westward. From his lessons, the link between comets and prognostication would survive and be most influential throughout the Middle Ages. Origen of Alexandria (ca. 185–253), an early defender of an evenhanded, allegorical interpretation of Revelation who lived a hundred years before Saint Augustine, suggested that the Star of Bethlehem, seen during Jesus' birth, was a comet or a meteor. He argued that, although such objects usually prognosticated the fall of dynasties and the outbreak of war, they might also signal an auspicious event. The early-fourteenth-century Italian master Giotto di Bondone, the first painter to truly animate Christian figures, seems to have shared this point of view; his 1304 *Adoration of the Magi* displays the Star of Bethlehem as a comet. In fact, his model was Halley's comet, which Giotto saw in 1301 (see figure 10 in insert).

Origen's optimistic take on comets was by far the exception. If Giotto indeed placed the comet as a positive sign, others interpreted it as a reminder of the forthcoming judgment. Chief among them was Thomas Aquinas (ca. 1225–1274), who not only discredited the cometary nature of the Star of Bethlehem (comets do not appear during daytime, was one of his arguments) but even attached doomsday significance to any comet: "On the *seventh* day all the stars, both planets and fixed stars, will throw out fiery tails like comets," he wrote.[12]

Not much was added to cometary theory or the discussion of other worlds during the Middle Ages. Only in the fifteenth century will we see a change, if not in formulating new ideas about comets, at least in making better measurements of their positions. Even though this increase of interest was triggered, not surprisingly, by astrology, the emphasis on accuracy is historically important.

A curious exception to the contagious allure of star-based prophecy is one work of Johannes Müller (1436–1476) of Königsberg, known as Regiomontanus. In his short treatise on comets known as *The Sixteen Problems on the Magnitude, Longitude, and True Location of Comets*, Regiomontanus explicitly avoided linking his ideas on how to measure distances and properties of comets to astrology, about which he also wrote at length. He was interested in obtaining actual numbers describing comets, such as the sizes of their head and tail, their distance to Earth and variable appearance, among others. His style was very modern in that it was concise, geometrical, and devoid of astrological speculation. Using a method for measuring distances of celestial objects known as parallax, Regiomontanus allegedly estimated that the comet of 1472 was at least 1,822 miles away.[13] Since he was off by quite a bit (the average distance to the Moon is roughly 240,000 miles), he found no reason—and did not want any— to challenge the Aristotelian belief that comets were sublunary. However, his pioneering use (or at least proposed use) of parallax marked a giant step toward a more precise astronomy, because only accurate measurements will force the revision of deeply ingrained preconceptions.

It is simple to see how parallax works: place a finger right in front of your nose, and look at it with the left eye only and then with the right eye only; you will see your finger "move" with respect to objects in the background. Now stretch your arm and do the same; your finger will still move, but much less. If you had a very long arm, it would move even less. The displacement angle of your finger with respect to a fixed object in the background is called parallax. The same is true for a celestial object: the closer it is to us, the more it will be displaced with respect to background constellations, when seen from different points on Earth's surface (see figure 11 in insert). The measured parallax of the comet of 1472 was 6 degrees. Its actual value is estimated with modern techniques at 3″—that is, 3 arc seconds, or $\frac{3}{3600}$ of a degree—clearly impossible to measure with the naked eye or the instruments of the time. The smallest angle we can discern with the naked eye is about 1′—an arc minute, or $\frac{1}{60}$ of a degree.

During the sixteenth century, two voices concerning comets and other celestial apparitions were heard side by side: while religious leaders and doomsday "prophets" fed even more superstitions and fear, a growing number of astronomers started paying more attention to data and less to astrological prognostication. This is not to say that debates about the physical nature of comets and the existence of other worlds were widespread in the 1500s; the reputation of comets as harbingers of doom was stronger than ever and mostly dwarfed any interest in a quantitative description of celestial events. The apparition of a comet in 1577 prompted Andreas Celichius, bishop of Altmark, a leading Lutheran reformer, to offer the following reflection on its nature and purpose:

> The thick smoke of human sins, rising every day, every hour, every moment, full of stench and horror, before the face of God, and becoming gradually so thick as to form a comet, with curled and plaited tresses, which at last is kindled by the hot and fiery anger of the Supreme Heavenly Judge.[14]

The Aristotelian mechanism for the formation of comets, exhalations rising from Earth ignited in the upper parts of the atmosphere, is transformed by apocalyptic rhetoric into evil rising from the souls of sinners and ignited by the wrath of God. In 1579, Andreas Dudith, a Hungarian cleric and humanist, replied that "if comets were caused by the sins of mortals, they would never be absent from the sky."[15] No doubt! The debate on the theological nature of comets is to be contrasted with the discovery, in the 1530s, that a comet's tail always points away from the Sun. Progress was slow but present.

Cosmic Boils

It was the great Danish astronomer Tycho Brahe (1546–1601) who delivered the first serious blow to the Aristotelian cosmic order. In 1543, Copernicus published his treatise placing the Sun at the center of the cosmos, the planets plastered to concentric spheres around it; as the spheres turned, so did the planets

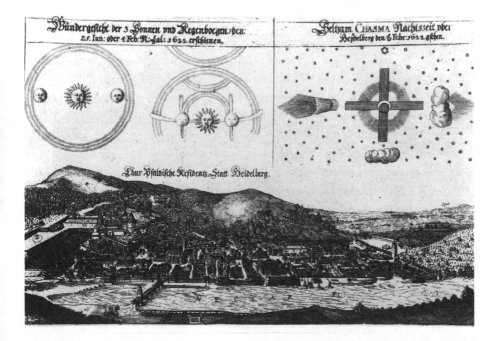

FIGURE 2: *Anonymous engraving depicting mysterious celestial phenomena, Heidelberg (1622). Left, two illustrations of multiple suns surrounded by rainbows; right, a comet and a cross-shaped apparition. All these phenomena were considered to be bad omens. (WOP-1. Courtesy Adler Planetarium and Astronomy Museum, Chicago.)*

At first sight, Copernicus's move was extremely daring and courageous for the times. However, in his mind, forward and backward thinking were deeply enmeshed: his Renaissance-inspired aesthetical ordering principle, of having a cosmic arrangement commensurate with the planetary orbital periods—Mercury, the shortest, first from the Sun, Saturn, the longest, last—was backed by an attempt to satisfy the Platonic ideal of celestial objects moving in circles with constant velocities.[16] The so-called Copernican revolution was to depend on the efforts of Copernicus's successors, namely, Tycho Brahe, Johannes Kepler, and Galileo Galilei, reaching its climax in the work of Isaac Newton during the late seventeenth century. As we shall see, these advances in the understanding of celestial mechanics initiated only a very gradual change in the widespread belief that comets and other unexpected celestial visitors were divine messengers. The rationally based description of their motions led to a reinterpretation of their purpose; instead of being mere tokens of God's wrath, they were seen as necessary agents of change, which shaped cosmic history and its cataclysmic transformations. Although doom became progressively more rationalized, the religious symbolism of comets and meteors remained very much alive, albeit disguised by a scientific language of quantitative precision.

On November 11, 1572, a "new star" flared in the constellation of Cassiopeia (the one that looks like a slanted W).* The eyes of most astronomers in Europe (and elsewhere) were riveted immediately on this new and extraordinary object. "What was it? Where was it?" investigators asked. The first answers to these two questions were, of course, Aristotelian: it is a sort of vapor condensation, hovering below the lunary sphere. Some (erroneous) parallax measurements seemed to confirm this. But Tycho was not convinced. He methodically observed the star until it faded out of sight in March. His conclusions were crystal clear: the new star was not a planet, since it did not move and, unlike planets, was far from the zodiac, the belt of twelve constellations—popular from horoscopes—where the planets parade. It was also not a comet, since comets move with respect to the fixed stars. Moreover, because the star failed to show any parallax, it was certainly farther away than the Moon. Contrary to Aristotelian belief, the skies could, after all, change. Tycho's observations, of unprecedented accuracy, were published in a book, *De Nova Stella* (The New Star), which also contained astrological interpretations of the event; after all, these were transitional times. Other astronomers, such as Thomas Digges in England and Michael Mästlin in Germany, agreed with Tycho, pub-

---

* Nowadays, we call this event a supernova explosion, which marks the end of a large star's life cycle. In part 3, we will discuss supernovae in some detail.

lishing their own observations, which they claimed offered support to the
Copernican system. Echoing Regiomontanus's modernity, they abstained from
including astrology in their treatises. A few decades later, the poet John
Donne's words illustrated how slowly astronomical discoveries percolated
through society,

> Who vagrant transitory comets sees,
>   Wonders because they're rare; but a new star
> Whose motion with the firmament agrees,
>   Is miracle; for, there no new things are.[17]

Sometimes a felicitous conspiracy of things happens at the right time to the
right people. The skies were quite generous to Tycho, who relentlessly pushed
forward his observational program based on accurate and methodical observations. On November 13, 1577, while casting a net to catch some fish before sunset, Tycho looked west and saw a "bright star which appeared as distinct as
Venus." At night, he measured the comet's tail to be over 21 degrees long, posi-
tioned at the "eighth house," which, according to Tycho's interpretation, signi-
fied forthcoming pestilence. Once it moved to the ninth house, the house of
religion, Tycho predicted the rise of new sects. While busy pondering the astrological significance of the celestial newcomer, Tycho also determined that, like
the new star in Cassiopeia, the comet had no parallax, being thus beyond the
lunary sphere. He estimated it to be at least 230 Earth radii away (Earth's radius
is 6,378 kilometers). Comets were finally let loose from the Aristotelian chains
that kept them anchored to Earth; they were now free to roam the cosmos and
shatter the crystalline spheres, which carried the planets in their orbits—and
anything else on their paths.

Tycho's conclusions that comets were far beyond the Moon did very little to
assuage the widespread belief in them as messengers of doom. Superstition
about the comet of 1680, more than one hundred years later, was as great as
ever. In 1684, John Edwards (1637–1716), a well-known Calvinist from Cambridge, wrote in his *Cometomania* that, apart from the comet's usual image as
harbingers of death, scarcity, famine, and the death of kings, these "funeral
torches to light Kings to their tombs" also had a healing function for the heavens, being like cosmic boils that collected corrupt matter and expelled it out, as
pus from a wound. Comets were

> a long Collection of corrupt and filthy matter . . . as in a Man's putrid
> Humours often gather in one part. . . . And these . . . excremental

humours breaking out, the Aether (like the Body of Man) is thereby kept
Sound and Hale. . . . The Sun and other Luminaries fare the better for the
expulsion of this gross stuff which would otherwise over-run them. . . . [18]

An opinion split had developed during the mid-1650s, mostly in England and
France, concerning the relation between comets and prognostication. Astrology
was decreasing in popularity, in particular among the ruling and educated
classes, who claimed it responsible for social unrest and the challenging of the
established order; rebels felt fortified by messages of the fall of kings, and prognostication often turned into self-fulfilling prophecy. Astrology and related sec-
ular prophecy were perceived as potential revolutionary weapons of great
importance; if God wrote his will in the stars and the stars said the king will
die, then he must die in order to fulfill God's will. So, let's go kill him! David
Gregory, of Oxford University, expressed this negative sentiment toward
astrology when he wrote in 1686:

> We prohibit Astrology to take place in our Astronomy, since it is sup-
> ported by no solid fundament, but stands on the utterly ridiculous opin-
> ions of certain people, opinions that are so framed as to promote the
> attempts of men tending to form factions.[19]

And yet, he allowed that comets might promote natural disasters if they ever
came into contact with Earth, their tails tainted with noxious vapors that could
poison our atmosphere and living beings. While most of the population still
nurtured the image of comets as supernatural portents, messengers of God's
wrath, the educated classes deemed this view "vulgar," accepting only their natural influences on Earth: like much else in England, opinions on comets and
celestial phenomena became related to class.

The Rationalization of Doom

Isaac Newton was certainly listening to the debate on whether comets were
portents, "cosmic boils," or just natural phenomena. The comet of 1680
appeared "twice," in November and December, and Newton carefully followed
its path, filling his notebook with comments and drawings. Although John
Flamsteed, the astronomer royal of England, insisted that the two apparitions

were the same comet at different points in its orbit, Newton initially rejected that claim. At the time he, like Edmond Halley, thought that comets obeyed laws distinct from those of planets; he leaned toward the rectilinear-path hypothesis proposed by Kepler earlier in the seventeenth century (see figure 12 in insert). However, by 1683, Newton, growing increasingly dissatisfied with the poor fit between theory and data, had completely changed his mind. During the fall of 1684, and three comets later (Halley's in 1682, followed by two others in 1683 and 1684), Newton put forward his *De Motu*, a short document in which he proposed that the motions of planets and comets were governed by a universal law of gravitation that varied with the inverse square of the distance between the celestial objects and the Sun. Comets, Newton surmised, moved in curved paths, some of them closed. This inspired Halley, a devoted follower of Newton's ideas, to conclude in 1705 that the 1682 comet was the same as that of 1531 and 1607, describing a highly elongated elliptical orbit with a period of 75.5 years. In 1687, upon Halley's persistent pressure, Newton finally finished his *Principia*, one of the greatest achievements of the human intellect. In successive editions, Newton would perfect his cometary theory with new observational evidence, such as Halley's, successfully predicting their orbital paths across the skies. The mystery of cometary orbits was finally solved.

After applying his theory to several different comets in book 3 of the *Principia*, Newton put forward a curious idea concerning the fate and purpose of comets:

> So, fixed stars, that have been gradually wasted by the light and vapors emitted from them for a long time, may be recruited by comets that fall upon them; and from this fresh supply of new fuel those old stars, acquiring new splendor, may pass for new stars. Of this kind are such fixed stars as appear on a sudden, and shine with wonderful brightness at first, and afterwards vanish by little and little. Such was the star which appeared in Cassiopeia's Chair.[20]

"New stars," such as the one observed by Tycho in 1572, were just old stars that got their fuel replenished by absorbing falling comets! The gravitational pull from stars would gradually attract orbiting comets, until they could not but fall on them. Newton then continued his musings on cosmic replenishment, extending his vision to planets:

> The vapors which arise from the sun, the fixed stars, and the tails of comets, may meet at last with, and fall into, the atmospheres of planets

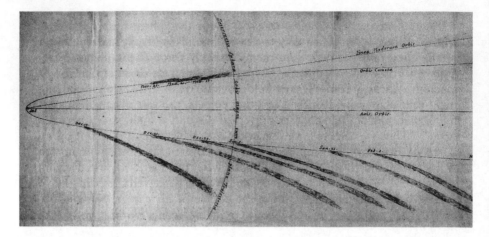

FIGURE 3: *Diagram depicting the orbit of the comet of 1680, from Newton's* Principia (1687). *The comet's elliptical orbit is approximated by a parabola in the vicinity of the Sun.*

by their gravity, and there be condensed and turned into water and humid spirits; and from thence, by slow heat, pass gradually into the form of salts, and sulphurs, and tinctures, and mud, and clay, and sand, and stones, and coral, and other terrestrial substances.[21]

The importance and beauty of Newton's scheme can scarcely be overestimated. His description of a constant ebbing and flowing of material substances across the heavens expressed an organic, alchemical vision of the world. Newton envisioned the cosmos as a web spun by wandering comets, which are responsible for the renewal of stars and planets and for the generation of substances that will ultimately support life. Moreover, he implicitly assumed a similar material composition of all planets, made of "terrestrial substances" obtained from stellar and cometary "vapors" cooked over slow heat (a reference to the slow burning of alchemical experiments): in Newton's vision, there was only one cosmic chemistry. But this is only half of the story.

The other half comes from Newton's belief in a teleological role for comets. Comets were God's tools, used for creating, renewing, and destroying worlds. In his mind, cosmic phenomena served a divine purpose. His physics explained motions on Earth and in the heavens, his alchemy the organic interconnectedness of the whole cosmos, and his theology justified the motions promoting these material exchanges as part of God's plan. It all fitted perfectly. There was no room here for cheap prophecy, which Newton deeply despised: "The folly of interpreters has been, to foretell times and things by this

Prophecy [Book of Revelation], as if God designed to make them Prophets. By this rashness they have not only exposed themselves, but brought Prophecy also into contempt."[22]

The marriage of physics with theology had the clear goal of justifying God's actions through rational mechanisms, in order to predict not *when* things would happen but *how* God operates in the world. Newton had no doubt that all the prophecies of Revelation would be fulfilled—some already had been, he argued—but he had no interest in predicting when. As we remarked before, for Newton the fulfillment of prophecy was evidence for God's presence in the world: "the event of things predicted many ages before will then be a convincing argument that the world is governed by Providence."[23] From this we can conclude that Newton saw the natural philosopher as a new kind of prophet—or better, the only kind of prophet—an interpreter of God's work as it was manifest in nature. In Newton's mind, science and religion were one indissoluble whole.

In keeping with the spirit of Revelation, Newton devised two mechanisms for the destruction of worlds by comets: direct impact—which Halley later took up—and falling into the Sun, promoting huge fires that would consume Earth and its living creatures. "He could not say when this comet [1680] would drop into the Sun . . . but whenever it did, it would so much increase the heat of the Sun, that this Earth would be burnt, and no animals in it could survive," wrote Newton's nephew-in-law John Conduitt, after a famous fireside chat they had in 1725.[24] In Newton's scheme of the world, history was punctuated by catastrophes promoted by collisions with comets through the agency of God, in what might be called a causal theology. Inadvertently, Newton redressed negative popular beliefs on comets and other celestial objects with the language of science; although there was no longer room for "local" forecasting, linking comets to the death of rulers or famines and plagues, the laws of nature allowed for the continued belief in comets as agents of doom, the ushers of a new age, just as prophesied in Revelation.

If Newton kept his thoughts mostly secret, revealing them only to a few select friends, Halley would be the spokesman of doom in the robes of the scientist. He had successfully demonstrated that Newton's physics explained the orbits of comets, predicting the return of those with closed orbits. To arrive at this conclusion, Halley had to painstakingly reconstruct the trajectories of comets from astronomical observations and compare them to predictions from Newton's theory, a truly amazing feat. But even before his discovery, Halley had argued that comets played a cataclysmic role in Earth's history—in particular, that they explained events described in the Old Testament. His favorite was a natural theory of the deluge, caused by "the casuall Choc of some transient body, such as a Comet or the like."[25] According to Halley, the collision would have abruptly tilted Earth's axis and diurnal rotation, the unbalance causing major flooding all over the planet. He went so far as to conjecture that an impact crater should exist in the Caspian Sea! As we will see later, impact craters are today the smoking guns of such cataclysmic events. However, influenced by a "secret friend" whose opinion he greatly respected (probably Newton), he pondered how Noah's Ark could have survived such a mess. By 1696, he had decided that the impact brought not the deluge but the end of a former world, happening thus before Creation; Halley believed, as did some pre-Socratics, in a succession of former worlds, each destroyed by an impact with a celestial body.

Halley closed his famous 1705 essay on comets by arguing for the possibility that such impacts could occur; as an example, he cited the comet of 1680, which passed very close to Earth. In a thinly disguised dramatic tone, he left open what the consequences of such impacts might be: "But what might be the Consequences of so near a [proximity]; or of a Contact; or, lastly, of a Shock of the Celestial Bodies, (which is by no means impossible to come to pass) I leave to be discuss'd by the Studious of Physical Matters."[26] Just in case, in another publication, Halley asks for God's help to avert such cataclysms: "May the great good God avert a shock or contact of such great Bodies moving with such forces . . . lest this most beautiful order of things be intirely destroyed and reduced into its antient chaos."[27]

The Newton-Halley ideas concerning impacts with celestial bodies provoked all sorts of responses, from church leaders to writers and social chroniclers. Jonathan Swift, in his *Gulliver's Travels* (1726), sarcastically translated this anxiety to the diminutive Laputans, who panicked at the approach of a doomsday comet within thirty-one years, the time predicted by Halley for the return of the 1682 comet. Preachers used it for their own purposes, urging sinners to repent while there was still time, now that comets had been *proven* to be God's instruments of apocalyptic doom: science validated prophecy. William Whiston, whom Newton nominated his successor to the Lucasian Chair of Mathematics at Cambridge, went as far as to suggest, in 1717, that comets were a perfect place for hell. Alternating spells of terrible cold (when far from the Sun) and heat (when near the Sun), comets were a "place of torment" for the sinful, doomed to circle the heavens throughout eternity. Following Whiston's lead, Cotton Mather, whom we met in chapter 2, preached that comets were former planets that God's wrath transformed into orbiting hells: comets were hell in the heavens.

The Genesis of Worlds

The mathematical elegance and predictive accuracy of Newtonian science, with its solution to the long-lasting mystery of celestial orbits and motions, gave natural philosophers a new sense of empowerment. A rational description of nature was now possible, based on the application of physical principles to natural phenomena. Nature was an open book, waiting to be explored by sagacious scientific minds. To many natural philosophers, there was no limit to what could be accomplished. This inflated confidence ushered in the rationalism of the eighteenth century, where every aspect of nature and life was believed to be explicable by the application of a set of general principles or laws to the world, society, and the individual. If, for Newton, God was a necessary presence in the cosmos, a sort of mechanic who kept celestial motions in check against perturbations in their orbits caused by gravitational attraction, for the new rationalists God's role was relegated to that of Creator. Once the material world had been created by divine power, its evolution unfolded under the strict jurisdiction of physical laws. The cosmos became a watch, and God the watchmaker. Ironically, the very precision of Newton's science undermined his demand for God's continuous presence in the world. The task of the natural philosopher was (and is), after discovering these laws, to apply them to the myriad phenomena of nature. In doing so, he was bringing himself closer to the mind of the Creator, his intelligence an atom of God's.

Newton left one fundamental question untouched: the origin of the planets or, more generally, of the solar system itself. In a letter to Richard Bentley, a chaplain to the bishop of Worcester interested in applying Newton's ideas to prove the existence of God, Newton confessed he had no idea why one single body should emit light at the center while other opaque bodies orbit around it. "I do not think [this arrangement] explicable by mere natural causes, but am forced to ascribe it to the counsel and contrivance of a voluntary Agent."[28] Later on, Newton attributed to the same agent the initial push that set the planets into their solar orbits. Newton's followers would be much more ambitious, and would try to understand the origin of the solar system simply in terms of mechanical principles.

The Frenchman René Descartes (1596–1650), who died when Newton was eight years old, used to say, "Give me matter and I will construct a world out of it."[29] Although lacking in a precise, quantitative formulation, Descartes's model of the cosmos was the first to appeal to a mechanistic description. He envisioned an infinite universe with an infinite number of stars surrounded by planets, as in our solar system. The stars were the centers of vortices of a diffuse, fluid matter, which filled all space, somewhat like the whirlpools around drains of an emptying bathtub, or in turbulent rivers. The circling motion of the vortices dragged the planets in their orbits. Descartes imagined a cosmos structured like a quilt, each piece housing a star and its surrounding vortex. Planets with moons had their own small vortices, the pattern being thus repeated on a smaller scale, an idea that reappeared later. According to Descartes, this ordered structure with its circular motions was the initial work of God, whose wisdom was manifest throughout nature. There were three different kinds of matter: the fluid matter of the vortices, the "speedy" matter that composed the Sun and stars, and the "coarse" matter that composed the planets. The rotating fluid vortex provided a separating force between the two other types of matter as a kind of cosmic centrifuge, the planetary matter being pulled out while the stellar matter somehow sank to the center. Anaxagoras's ideas found an echo here some two thousand years later. Although Newton criticized many of Descartes's ideas, including the existence of cosmic vortices, some of them surely influenced later models of solar system formation. More than anything, Descartes rendered plausible the possibility that the geometric arrangement of the cosmos was amenable to rational thought.

Immanuel Kant (1724–1804), whose great mind was imprisoned in a diminutive and frail body, is known mostly for his work on philosophy and metaphysics, and for an obsessive devotion to daily walks in a street now known as Philosopher's Walk, in his native Königsberg, Germany. It is said that Kant was so punctual that people set their clocks according to his walking schedule. In his youth, Kant was fascinated mainly by the arrangement of the cosmos and by Newtonian science. He was a firm believer in the deistic notion of God as Creator of the world and its physical laws and man as the decoder of its mechanics, as befit a true eighteenth-century rationalist. When he was thirty-one, still struggling to get a professional post, Kant wrote a treatise entitled *Universal Natural History and Theory of the Heavens* (1755), where he proposed a rather ambitious description of the arrangement of the cosmos, as well as of the origin of the solar system. Kant was so confident in the value of his work that he dedicated the treatise to Frederick the Great, king of Prussia, hoping the king's patronage would secure him an academic position. But fate was unkind to Kant, and his publisher went bankrupt just before distributing the freshly printed book. Most copies disappeared, and Frederick never got to see his. However, shortly thereafter, a review was published in a literary magazine and Kant's ideas started to be discussed in academic circles.

Kant's goal was to provide a compelling explanation for known facts regarding the solar system, on the basis of Newtonian mechanics and gravity.

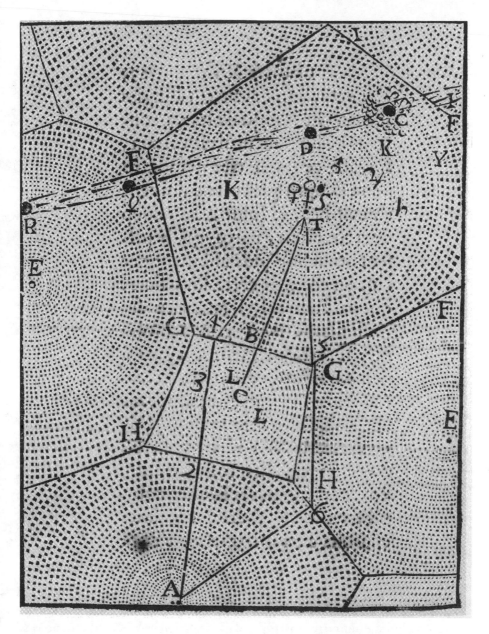

FIGURE 4: *Illustration of Descartes's vortices surrounding stars. Each vortex marks a polygonal domain, as in a sort of cosmic quilt. A comet travels from vortex to vortex, causing turbulence in the cosmic fluid. (René Descartes,* Le Monde *[Paris 1664].) (Reproduced from Sara Schechner Genuth,* Comets, Popular Culture, and the Birth of Modern Cosmology *[Princeton: Princeton University Press, 1997], p. 112.)*

That is, he wanted to show that these facts are a consequence of a dynamic formation process resulting from the motions of matter under the influence of various forces. Here are some observational facts pertaining to the solar system that Kant set out to explain: (a) The planetary orbits lie nearly on a plane intersecting the Sun's equator. That is, the solar system is quite flat. More precisely, the "thickness" of the disk is about $\frac{1}{50}$ of the diameter of Pluto's orbit. (Pluto was not known in Kant's time; but the flattened shape of the solar system was.) One can roughly picture this as a laser disk bisecting a Ping-Pong ball (not to scale!), the planets being little dots on the laser disk moving in orbits around the center. Comets, however, depart from this rule, having orbits at arbitrary inclinations with respect to the orbital plane of the planets. (b) All planets go around the Sun in the same counterclockwise direction as seen from above Earth's North Pole. This is also the same direction in which the Sun rotates about its axis.

In 1751, Kant became aware of ideas from the Englishman Thomas Wright (1711–1786), the first to propose a specific shape for the Milky Way. According to Wright, the Milky Way was an infinite collection of stars flattened between two infinite planes, somewhat like a cosmic sandwich. Kant conjectured that there was a deep connection between Wright's model of the Milky Way and the flattened shape of the solar system: both were formed in the same way, by the same physical processes, the Milky Way and its stars being just a larger version of the Sun and its planets. In fact, Kant proceeded to propose that the many "nebulae" that had recently been observed were merely collections of stars like the Milky Way, something that would be confirmed only in 1924, by the American astronomer Edwin Hubble. (Some were indeed galaxies, while others were gas clouds.) In Kant's view, the universe was filled with galaxies, each composed of innumerable stars, each star surrounded by its court of planets, and some planets surrounded by their court of moons. The same pattern repeated itself from the largest distances in the cosmos down to Earth and its Moon. Kant saw this magnificent cosmic symmetry as a clear manifestation of God's design.

The structure being set, it was necessary to devise a mechanical principle to explain how these shapes emerged in time. Kant assumed that, in the beginning, the infinite universe was filled with scattered particles, the basic constituents of all matter. He paid tribute to the atomists and their primordial chaos, writing that the ideas of Lucretius—whom he dearly loved—and of his predecessors Epicurus, Democritus, and Leucippus, "had much resemblance" to his own.[30] He revived the atomists' (and in spirit, if not in detail, Descartes's) general idea of vortices, arguing that they arose from the competition between an inward fall—present in Epicurus's system and explained by Newton's gravitational force—and a sideways motion, which he erroneously attributed to a repulsive force, as in the "elasticity of vapors, the effluences of strong smelling bodies, and the diffusion of all spirituous matters."[31] However fond of the atomists' physics, Kant did criticize their atheism, claiming that a universe without God's counsel would never have ordered from its initial chaos.

According to Kant, initial overdensities in the primordial chaotic matter started to attract more matter, in a simple infall motion determined by gravity. As more matter accreted, the "repulsive" forces among the particles pushed them sideways, generating a general circular motion, or vortex. While larger quantities of particles fell to the center, forming the Sun, smaller swirling particles in the vortex combined to form the planets. Kant believed that the material composition of the vortex was such that the closer to the center, the denser and more massive the particles. Thus, as one moved away from the center, the planets had larger quantities of lighter particles, which had more variations in their motions. As a consequence, the eccentricity of the planetary orbits, that is, their deviations from a perfect circle, increased with their distance from the Sun. Comets, the outermost inhabitants of the solar system, were the ones with the lightest components and thus the most eccentric orbits. Kant even conjectured

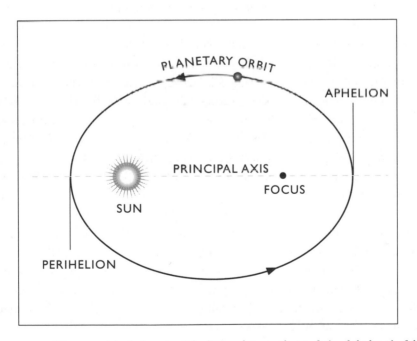

FIGURE 5: *The eccentricity is the ratio of the distance between the two foci and the length of the larger axis of the ellipse. For a circle, the distance between the two foci is zero and so is its eccentricity.*

that there would be other, more eccentric planets beyond Saturn. He was, of course, dead right about the existence of other planets, but dead wrong about their increasing eccentricities, as you can verify from table 1. (Note, in particular, that Neptune has a nearly circular orbit.)

PLANET	ECCENTRICITY
Mercury	0.206
Venus	0.007
Earth	0.017
Mars	0.093
Jupiter	0.048
Saturn	0.054
Uranus	0.047
Neptune	0.009
Pluto	0.249

TABLE 1: *Eccentricities of solar system planets.*

Kant's universe was largely built out of his uncanny intuition. His mechanism for the formation of the solar system followed the logical application of physical principles, which, as we have briefly seen, had a few basic flaws. As a consequence, some of his ideas were far-fetched, such as his notion of the increasing eccentricities of the outer planets, or his belief in an absolute center of the cosmos, to which all other secondary centers were hierarchically related. To Kant's credit, he was the first to admit that there were very few quantitative calculations in his work, although he clearly expected that his ideas would be confirmed by a more precise mathematical formulation. And, as we will soon see, some were. Quantitative precision aside, Kant's rhetoric expressed his passion for nature and his firm belief in the power of reason to face its most puzzling mysteries. His eloquence reached a climax when he speculated about the eternal cycles of creation and destruction that punctuate the vast cosmic distances. The same forces that create worlds out of inanimate matter will eventually cause their destruction, as, in the "fullness of time," they collapse into the massive center of their orbits. According to Kant, once time has run its course (a rather vague statement), a gradual falling of planetary and cometary matter into the central sun will trigger a "conflagration," which will rescatter their constituent particles through the vastness of space they once occupied. From this renewed chaos, small overdensities will again accrete more neighboring matter, and a new sun will be born, together with its court of planets. Kant believed these beautiful cycles of creation and destruction to be the very essence of nature, filling the eternity of time. Only God and the eternal soul were above these material processes of birth and death:

> When we follow this Phoenix of Nature, which burns itself only in order to revive again in restored youth from its ashes, through all the infinity of times and spaces . . . with what reverence the soul regards even its own being, when it considers that it is destined to survive all these transformations![32]

Pierre Simon de Laplace (1749–1827), a brilliant mathematician and physicist, took over where Kant left off. His approach to science was quite different from Kant's, being based on a solid mathematical description of physical phenomena. Apparently, he was not even aware of Kant's work. By a refined application of Newtonian mechanics to the motions of Jupiter and Saturn, Laplace was able to show that their mutual gravitational attraction would not affect their orbits around the Sun, hinting at the long-term stability of the solar system; even though planets tug at each other, there was no danger of their running off track. He used this result to argue for the safety of life on Earth, at least from the point of view of orbital instabilities. As a consequence, if the solar system could regulate itself through its own motions, there was no need to invoke God's constant presence, as Newton did; to Laplace, God was not a player in celestial mechanics. His stability analysis relied on a crucial physical concept, that of the conservation of angular momentum. Since angular momentum will be a recurring theme from here on, let us take a moment to review the basic ideas behind it.

We start by first considering linear momentum, which is defined simply as the product of a body's mass and its velocity. So, a bicycle moving on a straight line at 20 miles per hour and an eighteen-wheeler at 20 miles per hour may have the same velocity, but have very different momenta, because of their large mass difference. Linear momentum is a measure of an object's tendency to remain in its state of linear motion; brakes have to work much harder to stop an eighteen-wheeler than a bicycle. Descartes was the first to clearly suggest that, in the absence of friction or other losses, the total linear momentum is conserved when two or more bodies interact. That is, the sum of all the individual momenta is the same before and after a collision. You can visualize this in collisions involving billiard balls, as they hit each other and career with different velocities in different directions; although the conservation is not perfect, because of friction and thermal losses (for example, the fact that you hear the collision means that some energy is being lost during the shock), it's pretty close for our purposes. An even better example is the game of air hockey, where both friction with the table and the rolling of the balls are largely eliminated, improving momentum conservation.

Angular momentum is an extension of this idea for motions involving rotation. Imagine you are spinning a stone at the end of a string: the angular momentum of the stone is given by the product of its mass, its angular velocity (the rate at which it is rotating, say, 30 degrees per second), and the square of its distance from the center. Thus, the longer the string or the angular velocity of the stone, the larger its angular momentum, a measure of the tendency of a body to remain in circular motion. That, in the absence of friction, angular momentum is conserved is a concept we are all familiar with: when an ice-skater does her pirouettes, she usually starts with her arms outstretched. This sets her initial angular momentum, which depends on her arm's length. What happens as she brings her arms in? She spins faster and faster, until she reaches her maximum angular speed with her arms folded on her chest, to the enthusiastic applause of the crowd. This speeding up is easily seen as a consequence of angular momentum conservation: since the total angular momentum is conserved, it must be the same at the beginning, as the skater spins with her arms stretched out, and at the end, with her arms folded. Her mass staying the same, the only way to keep angular momentum constant as her arms retreat (smaller distance to the center) is to increase her angular velocity. If you have never seen a pirouetting ice-skater, try the following experiment: find a rotating stool, and spin sitting on it with your arms stretched out. While still spinning, bring them closer to your chest and notice how you spin faster. Voilà! You have demonstrated the conservation of angular momentum.

Returning to Laplace—he believed that the ordered arrangement of the solar system, with the planets spinning in tandem within a narrow plane and in the same direction as the Sun around its axis, was the result not of chance but of solid physical principles. He proposed the so-called nebular hypothesis, according to which the Sun's atmosphere once extended all the way toward where today we find the outermost planet (in Laplace's days, Uranus). So, unlike Kant, Laplace started not with some primordial universal chaos but with the primitive Sun surrounded by a diffuse, fluidlike, and fiery atmosphere. In fact, he refrained from applying his ideas to other stars or to the cosmos as a whole, limiting himself to the solar system. He further assumed this vaporous mass to be slowly rotating around the Sun, thus establishing its initial total angular momentum. However, he did not venture to explain the origin of the vaporous mass or its angular momentum. According to Laplace, as the solar primordial atmosphere was pulled by its own gravity toward the center, it also gradually cooled off (modern physics predicts the opposite for a contracting gas cloud, a serious problem with Laplace's ideas). Because of the conservation of angular momentum, as it became smaller, the contracting gaseous nebula rotated faster, like a cosmic ice-skater. The combination of contraction with faster rotation caused the nebula to bulge around its equator, becoming flatter the faster it spun. Rotating pizza dough does the same.

Laplace conjectured that, at a certain point during the contraction, the outer portion of this rotating mass broke off from the rest, as its attraction to the central Sun was balanced by the centrifugal force caused by its motion. This created a ring of matter detached from the rest of the further contracting mass, until another, smaller ring was created and so on (see figure 13 in insert). Within each of these concentric rings, material particles slowly collided and coalesced into large massive bodies, which grew to become the present planets of the solar system. Variations in the materials running around the ring, that is, rocky versus volatile substances, explained the differences in the material composition of the planets. This larger pattern repeated itself on a smaller scale: each protoplanet, at a certain point during its evolution, also acquired rings of matter around it. These, upon further cooling, became the planet's satellites or, sometimes, diffuse rings made of smaller separate bodies, as in Saturn.

Laplace's influence as a scientist was very widespread: "Before the imposing greatness of Laplace all bow" was the watchword around scientific circles in the early nineteenth century.[33] Contrary to general belief, science is not immune to power-based opinions: fashions and trends abound! However, its advantage over other areas of human activity is that, sooner or later, opinions supported by arguments based on power and not on substance are overthrown: a wrong hypothesis, no matter who proposed it, will hold for only so long (unless, of course, it is so far removed from the reality probed by experiments and observations that it can survive for a long time). Slowly but surely, several questions emerged concerning the plausibility of Laplace's scenario. Scientists, especially outside of France, questioned the fact that he did not offer a mechanism for the origin of the Sun and its atmosphere, or their overall rotation. Also, at that time it was known that at least two of the satellites of Uranus orbited backward, that is, had a retrograde orbital motion, which contradicted the uniform rotation implied by Laplace's hypothesis. Furthermore, it was not clear how matter in the rings coalesced into very large planets, it being more plausible to assume they would at most form small bodies spread in a belt, such as the one found between Mars and Jupiter, the so-called asteroid belt. Comets were also a problem: since they have very eccentric orbits, making arbitrary angles with the orbital plane of the planets, Laplace was forced to conjecture that they did not belong to the solar system, being small nebulae with dense kernels crisscrossing the cosmos from star to star. Finally, the Sun rotates more slowly than a straight application of conservation of angular momentum would predict.

Given the importance of linking the development of astronomy with religious and popular culture, before we move on to modern theories of the genesis of worlds, it is worth looking a bit further into Laplace's views on comets. Since comets were free-roaming cosmic travelers, they could sometimes get caught by the Sun's gravity, like a fish by a net, occasionally locking into the periodic orbits established by Halley. Laplace used his celestial mechanics as a weapon against the irrational fear of comets still rampant among large sectors of the population. Or such was his intention. He showed that the tails of comets were so rarefied that their total masses were smaller than that of a small hill; their effects upon brushing Earth were completely negligible. (Nevertheless, as we have seen, terror still spread like the plague when Halley's comet's tail brushed Earth in 1910, carrying its "deadly" cyanogen.) Laplace praised "the light of science [that] has dissipated the vain terrors which comets, eclipses, and many other phenomena inspired in the ages of ignorance."[34] He also argued that, although possible, collisions with comets were extremely rare. This being the case, his scientific honesty forced him to add, "The small probability of this circumstance may, by accumulating during a long succession of ages, become very great."[35] That is, cosmic collisions, science was confirming, do happen. From the perspective of popular culture and religion, Laplace and his colleagues appeared to be confirming doomsday prophecies, wearing the robes of rational science: the temptation to prognosticate, even if justified by a sense of scientific integrity, was just too great. Laplace went on to describe the effects of such a collision, without trying to minimize its apocalyptic consequences:

> The [Earth's] axis and motion of rotation changed, the waters . . . precipitate themselves towards the new equator; the greater part of men and animals drowned in a universal deluge, or destroyed by the violence of the shock given to the terrestrial globe; whole species destroyed; all the monuments of human industry reversed.[36]

Here is a short parable on power and fear, which some of you may identify with. You go to a doctor with a red spot on your skin. After a cursory examination, he says, his eyes gleaming with the look of those who can see death within life: "Listen, there is a very small chance that your skin rash is actually an early symptom of a devastating flesh-eating disease, that could basically devour you within a couple of days." "Wow, Doc, thanks! I would never have thought of it," you say, sweat running down your forehead. I am sure you will not sleep well for the next two nights. In the midst of the ensuing panic, the cautionary words "a very small chance" quickly lose their meaning. Fear quickly blurs sensibility. After an agonizing week, your rash disappears and you realize you were not the one-in-a-million case. But you could have been, right?

Laplace's ambiguity in regard to the destructive powers of comets—rare but fatal—reminds me of a 1967 Roman Polanski movie, *The Fearless Vampire Killers*. It tells the story of a professor who sets off for Transylvania with his assistant (played by Polanski) to prove his theories about the existence of vampires and, while he's at it, to exterminate the remaining ones. After encountering quite a few of them, including a Jewish vampire who laughs defiantly when the professor threatens him with the cross ("Oy, you got the wrong vampire!"), his assistant falls prey to the innkeeper's daughter-turned-vampire, played by Sharon Tate. The professor knows nothing of this and, after causing major havoc in the castle where the vampires hide, escapes with his contaminated assistant and his vampire girlfriend back to western Europe. The movie closes with the line "And thanks to Professor Abromsius, the evil he sought to exterminate will spread throughout the world." Laplace's rhetoric ended up doing the same.

The Genesis of Worlds, circa A.D. 2001

In the two hundred or so years since Laplace developed his nebular hypothesis, astronomers have proposed a variety of ideas that either tried to fill the gaps of his scenario or to set forth very different ones. There have been catastrophic formation models, where planets are chunks of matter yanked from the Sun by the near-encounter with another star; giant protoplanet scenarios, where the planets themselves are the product of condensing nebular material, clouds broken off from the giant nebula that became the solar system; and planetesimal scenarios, where the planets were formed by a gradual accumulation of material. This last scenario is the one widely accepted today.

In order to explain the formation of the solar system, we must start at much larger spatial scales and a long time ago, with the formation of our own galaxy. (Figure 14 in insert shows a spiral galaxy similar to ours.) Present estimates place the general era of galaxy formation around twelve billion years ago, when the universe itself was at the tender age of one or two billion years. (For now, we will take the age of the universe to be fourteen billion years.) In older versions of galaxy formation scenarios (the top-down scenarios), the Milky Way emerged from the contraction of a giant nebula—as Kant suggested—with a mass of a few hundred billion suns. Its initial composition was very simple, mostly hydrogen and helium, the two predominant chemical elements in the universe, appearing in the proportion of 3:1; for each four atoms, three were hydrogen and one helium. Recent studies suggest instead that large galaxies, such as the Milky Way, formed through a succession of mergers of smaller nebulae, in a sort of cosmic cannibalism. In any case, the chemistry of the merging nebulae follows the 3:1 rule. The details of galaxy formation remain very much a topic of debate in the astronomical community.

Small overdensities, regions where the amount of matter in a given volume exceeded the average value across the nebula, like clots in bread dough, triggered the contraction of hydrogen/helium clouds because of their own self-gravity. The more these regions contracted, the hotter and denser they became. After a few million years, the temperature at the core of the denser regions was so high that a process known as nuclear fusion started to occur: nuclei of hydrogen atoms (one proton) would fuse into nuclei of helium atoms (two protons and two neutrons), liberating enormous amounts of energy. The largest of these early stars, thousands of times more massive than the Sun, started to shine with tremendous fury, their existence hanging on the balance between its own gravitational contraction and the energy liberated outward by nuclear fusion, implosion versus explosion. But gravity never rests. In the end, it always wins, and the huge stars collapsed over themselves. As the shrinking outer regions of the star hit the very dense core, they recoiled with a bang, ejecting large amounts of matter across the interstellar medium. In part 3, we will discuss the life cycles of stars in detail. For now, it is only important to keep in mind that these supermassive stars had relatively short lives, spewing huge amounts of matter into the primordial interstellar medium as they convulsed into oblivion. About 3 percent of this matter was composed of heavier chemical elements, such as carbon, oxygen, and nitrogen, cooked by the continuation of nuclear buildup, which had started with hydrogen going into helium. After billions of years, these short-lived supermassive stars seeded the early galaxy with tiny grains of matter, which were going to be key ingredients during the formation of our solar system.

As several nebulae merged on their way to becoming the Milky Way, matter accumulated at the galactic center and its beautiful spiral-arm structure became more and more defined.* Scattered across our proto-galaxy, smaller overdense gas clouds of different sizes, shapes, masses, and temperatures orbited around the center, colliding with or passing near each other, while being seeded by grains spewed by the dying supermassive stars in their neighborhoods. A bit over 4.5 billion years ago, one of these clouds began to contract, possibly triggered by a near collision with another cloud, by its passage through an overdense region of a galactic arm, or by a nearby stellar explosion. This cloud of hydrogen and helium, sprinkled with the heavy-element seeds spread throughout the galaxy, would become our solar system.

As the rotating solar nebula contracted and flattened, the temperature at its core started to increase, as happens with supermassive stars. However, since the nebula was considerably less massive, the hydrogen fusion process there was not as furious; after a very luminous initial stage that lasted a few dozen million years, the core cooled off and settled into a more gentle hydrogen burning pace, which would sustain itself for about ten billion years. The Sun is in its middle age now. During its early stage, it looked like an overheated star surrounded by a fairly flat gaseous disk, sprinkled with heavy-element dust and gas left over from previous stellar explosions (see figure 15 in insert). As it cooled down, some of the matter in the gaseous disk condensed into solid grains, whose sizes ranged from microns (one millionth of a meter) to millimeters. Since different substances solidify at different temperatures, specific kinds of solid grains were more common at different distances from the Sun, from the inner regions (temperatures of a few thousand degrees) to the cooler outskirts of the disk (a few degrees above absolute zero). We are familiar with the fact that different substances freeze at different temperatures; if you place a bottle of water in the freezer, the water will turn to ice, but vodka will become more viscous, without freezing. Likewise, different substances will solidify at varying distances from the Sun, as represented in the figure below. It is this distribution of different solid materials across the solar nebula that determines the material composition of the planets.

Once the distribution of grains is established, the process of planet buildup occurs in three stages. First, the small grains act as seeds for the condensation of more grains around them. This is what happens with rain as well; dust or soot

---

* Part of the debate concerning the mechanism of galaxy formation is related to the apparent fragility of spiral galaxies, such as the Milky Way, against violent mergers. For us, it is sufficient to assume that whatever the details of galaxy formation, overdensities that evolved to become solar systems were present about five billion years ago.

in the air attracts water vapor molecules and helps them release heat. As the clusters of molecules cool, they condense into large liquid droplets that fall to the ground—that is, it rains. This is why we sometimes see planes "sprinkling" clouds with salts, in the hope that it will speed up the condensation process and produce the much needed rainfall. Back to planet formation—solid grains of material grow to become small clumps of matter, grains sticking to grains, as snow sticks to snowballs. The larger the clumps, the larger their surface area and the more grains stick to them. This process of accretion continues for a while, until the clumps grow into objects a few hundred kilometers across. At this point, the clumps are massive enough for their gravity to start helping the collection of nearby material; the process speeds up further, and the clumps grow to the size of small moons, or planetesimals, completing the first stage of planetary buildup.

During the second stage of planet formation, lasting about one hundred million years, the combination of motion and mutual gravitational attraction caused all sorts of collisions among the millions of planetesimals orbiting the central mass. In this treacherous environment, the rich became richer, as small planetesimals were either destroyed or absorbed onto larger ones. Computer

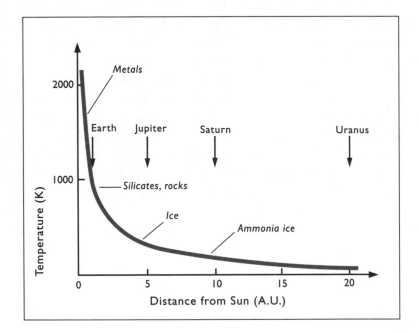

FIGURE 6: *Different substances solidify at different distances from the Sun, as is indicated in the diagram. At large distances (Jupiter and up), there is a marked absence of rocky materials.*

simulations show that this cleanup process also made the orbits of the larger planetesimals more circular. Essentially, the objects that survived this process of collisions and bombardments became the nine planets of the solar system. There is also the asteroid belt, between Mars and Jupiter, a collection of millions of planetesimals that never formed into a planet, from huge Ceres, with a diameter of 913 kilometers, to roughly 28 million rocks larger than a football field. The reason for this failure is probably Jupiter's huge gravitational pull, which made their collisions too violent, preventing the sticking needed for efficient growth.

The outer planets, from Jupiter on to Neptune (Pluto is a special case), entered a third stage of planetary buildup, sweeping up large quantities of gas present in the outer regions of the solar nebula. We know that these planets are very different in composition from the "terrestrial" planets, that is, Mercury, Venus, Earth, and Mars: inner planets are rocky, whereas outer planets are gaseous. We can understand these differences if we briefly revisit the formation process, focusing on the outer regions of the solar nebula. Since the temperatures there were lower, more gases could solidify, and this abundance of frozen matter jump-started the accretion process forming the planetesimals: the outer planets grew larger because their surrounding medium was richer. This buildup material was mainly in the form of compounds of water vapor (H_2O), ammonia (NH_3), and methane (CH_4), which could not solidify in the warmer regions closer to the Sun (H stands for hydrogen, O for oxygen, N for nitrogen, and C for carbon). The outer planetesimals grew large enough to accrete the hydrogen-rich nebular gas around them, until they reached their enormous sizes.

The icy planetesimals left over from this accretion process were flung either into the inner regions of the solar system or out, by the enormous gravitational pull of the gaseous giants. A vast cloud of these small icy objects exists in the outskirts of the solar system—the Oort cloud, named after the Dutch astronomer Jan Oort, who first proposed its existence in the 1950s. In fact, this is the largest of two cloud nurseries of comets, where these objects, icy debris left over from the formation of the outer planets, move in their orbits around the Sun. The other, known as the Kuiper belt, is located outside Neptune's orbit. Once in a while, as a result of rare collisions, or close encounters with neighbors or the proximity of a nearby star or a gas cloud, an icy ball becomes destabilized and is sucked into the inner solar system: a comet is born. As the comet approaches the Sun, increasing temperatures warm up its materials and start vaporizing them, one by one. The point at which a given material burns is determined by the ambient temperature, that is, by the comet's distance from the Sun. The different compositions of the comets account for the differences in colors observed in their tails, which are nothing more than vaporized material pushed back by the pressure coming from the solar wind, an outward flow of fast-moving charged particles produced by the Sun.

The modern account of the formation of the solar system, based on the condensation of the primordial solar nebula aided by interstellar dust, matches most, but not all, of its observed properties. It does not explain why certain planets (Venus, Uranus, and Pluto) rotate around their axis opposite to the Sun's rotation; or why Uranus's axis of rotation is so tilted—almost 90 degrees—with respect to the plane of the ecliptic; or why Mars has such a rarefied atmosphere; or why Earth has so much water; and so on. These odd properties are explained by invoking processes that occurred mostly after the formation of the planets. More to the point, as these properties depart from the apparent order of the solar system, their causes are attributed to catastrophic phenomena of various sorts, from meteoric and cometary bombardments to collisions with satellites or large asteroids. Thus, in order to complete our modern description of the solar system, we must come full circle to its catastrophic side.

Our present understanding of solar system dynamics incorporates cataclysmic phenomena in its everyday language. It leads to the inescapable conclusion that destruction is part of creation; that cosmic events unfold irrespective of any moral judgment on our part. It is neither right nor wrong for a comet to collide with a planet: it just happens when their paths meet in the vastness of the cosmos. Throughout the ages, religions have charted our beliefs, inspired by the awesome forces of nature, seen as manifestations of divine power; for example, the rains that gave us food also caused devastating floods, comets may herald the birth of Jesus or the impending punishment of sinners. To a large extent, modern science has changed our attitude toward natural phenomena; what was once explained away by divine causes is now seen as a consequence of well-understood physical causes. Contrary to widespread popular belief, this does not mean that modern science took away the magical beauty of celestial phenomena. Quite the opposite—our understanding of the workings of the cosmos only adds to our appreciation of how awe-inspiring it truly is. Moreover, it teaches us that, if we are wise enough to monitor the comings and goings of our racing celestial neighbors, so that we can act in our defense if needed, we will be able to remain on this planet for a very long time. Assuming, of course, that we will also keep an eye on ourselves, and on our own destructive power. The alternative is to join the cycle of creation and destruction that perpetuates itself across the universe, and burn in a "conflagration," as countless other planets have and will.

Religious prophecies translated our fear of the skies into a yearning for eternal salvation; falling stars marked the end of history and the beginning of eternity. Science confirmed the possibility that the destructive side of these prophecies may very well be fulfilled one day. But it also gives us, perhaps paradoxically, hope for salvation. To see why, we turn to a discussion of catastrophism in the solar system, the meeting ground of apocalyptic fears and apocalyptic science.

CHAPTER 4

Impact!

The Committee believes that it is imperative that the detection rate of Earth-orbit-crossing asteroids must be increased substantially, and that the means to destroy or alter the orbits of asteroids when they threaten collision should be defined and agreed upon internationally.

— U.S. HOUSE OF REPRESENTATIVES,
NASA MULTIYEAR AUTHORIZATION ACT OF 1990

The early history of the solar system was marked by violent encounters between the growing planets and smaller planetesimals. This was the "cleaning house" era, which lasted until 3.9 billion years ago, when young planets traveled in an environment filled with debris left over from the accretion process. Their satellites, of course, were not spared the bombardment. The surface of the Moon, pocked with craters of all sizes (over thirty thousand of them), offers testimony to its violent beginnings. Several satellites of the giant planets display the same tortured surfaces. Venus is covered by such a thick blanket of noxious fumes that direct viewing of its surface is impossible. However, radar images by the *Magellan* spacecraft exposed over nine hundred craters, ranging from a few kilometers in diameter to the giant Mead crater, some 280 kilometers. Mercury's surface is likewise covered with craters. Earth surely did not escape this early bombardment, although these ancient craters have mostly been erased by intense erosion promoted by continuous atmospheric and seismic activity. Of the few known craters, most are fairly recent; for the celestial bombardment, if less intense now than during the solar system's first billion years, is still very much a disturbing reality. An example is Meteor crater in northern Arizona, which has a diameter of 1.2 kilometers and a depth of 200 meters (see figure 16 in insert). The main reason we can still see this particular crater is its very young age, about fifty thousand years. Collisions with asteroids and comets are not relegated to the very distant past. Although rarer now, they can happen anytime. It was only in the 1960s that Eugene Shoemaker, a leading authority on impact geophysics, proved convincingly that the Arizona crater was caused by the impact of an iron-rich meteorite about 50 meters in diameter. The alternative explanation at the time held that the crater was the result of some violent volcanic activity. Shoemaker and his collaborators found samples of high-pressure glassy rock and deformed geological structures produced by the tremendous violence of the impact, and that finding put the debate to rest.

Picture a rock the size of a fifteen-story building hitting the ground at 11 kilometers per second (roughly 25,000 miles per hour). Impact physics is about momentum and energy transfer. Both of them must be the same before and after the impact: what the celestial offender gives, Earth receives. Because momentum is mass times velocity, and Earth's mass is so much larger than any asteroid or comet (roughly three billion times larger for a 5-kilometer-wide asteroid), the change in Earth's momentum as a result of a collision is perfectly negligible: an impact will not knock Earth out of its orbit, just as you cannot move a truck by hitting it with a baseball. With energy, however, things are very different.

The amount of energy before impact—the energy of motion, or *kinetic energy*, of the incoming projectile—must be equal to the energy after impact, that is, the energy dissipated on the ground and atmosphere around it. Now, the kinetic energy is proportional to the mass times the *square* of the projectile's velocity. It is the *square* of the velocity that makes all the difference, especially when the velocities are of tens of thousands of miles per hour. You can verify this by throwing rocks of about the same size into a pond, varying their velocity at each throw and watching the results. The harder you throw the rocks, the more of a disturbance you cause on the surface of the pond (see figure 17 in insert). You see the kinetic energy of the rock being dissipated by concentric water waves propagating outward from the point of impact; you see the recoiling water rising up into the air and falling back again, water drops spreading around a large area; and, for very hard throws, you see the rock penetrate into the bottom of the pond.

Although the velocity plays a key role in determining the damage due to an impact, the mass is also clearly important. Everybody laughs when little five-year-old Johnny does his cannonball dive on the pool, but when 200-pound Uncle Louie does the same, nobody finds it very funny. Earth is sprinkled daily with over a ton of micrometeorites the size of a grain of sand or smaller. They are the shooting stars, streaking the night skies with a feeble light, as they are vaporized by friction with the atmosphere. When falling rocks reach the size of a fist or a bit larger, all the atmosphere can do is slow them down, and an impact occurs. These are the meteorites we see in museums and planetariums. As for little Johnny and Uncle Louie, it should be quite clear which of the two has a greater chance of hitting the bottom of the pool.

Back to Meteor crater. Knowing the velocity and size of the rock, we can estimate the impact energy to have been equivalent to the simultaneous collision of about one billion 20-ton trucks moving at 100 miles per hour. Since we cannot pack one billion trucks into 50 meters (the estimated size of the rock), we can use another analogy; the impact energy was equivalent to the detonation of 20 to 40 megatons of TNT (mega = million). Since a large hydrogen bomb deploys about 1 megaton of TNT, the impact that forged Meteor crater is equivalent to dozens of hydrogen bombs exploding together (without the radioactivity, of course). The ground around the impact was instantly vaporized, together with most of the asteroid. About 175 megatons of rock were lifted by the impact, only to rain down over an area of about 9.5 kilometers from ground zero.[1] The collision created a shock wave called an *air blast*, with winds of over 1,000 kilometers per hour. Even as far as 40 kilometers from the impact site, winds were still of hurricane force. Fossils indicate that at the time of impact, during the last ice age, the area was populated with mammoths, mastodons, giant ground sloths, and other huge mammals. It is hard to imagine that any of these animals could have survived within a radius of at least 20 kilometers. If a similar impact were to occur in a metropolitan region today, millions of people would die in a flash. And this is only one of the little guys.

Widespread geological surveys from land and space during the past two decades led to the discovery of several impact craters. The total now stands at about 150 and is growing by 5 or so a year. The crater sizes and times of impact vary widely. The Haviland crater, in Kansas, and the Sobolev crater, in Russia, are only about one thousand years old, whereas the Vredefort crater, in South Africa, is two billion years old. The Haviland crater is 15 meters wide, whereas a huge crater in Sudbury, Ontario, could potentially be 250 kilometers wide. The conclusion is clear: Earth has been bombarded by large objects in the past and not so distant past, just like all other bodies of the solar system. Scientists use expressions such as "cosmic pinball," "cosmic shooting gallery," "target Earth," and so forth to illustrate the fact that collisions are an integral part of life in the solar system. Doomsday scenarios, once part of religious discourse, are now vested with scientific precision. We now know that a 10-kilometer-wide object fell over the Yucatán peninsula in the Gulf of Mexico sixty-five million years ago, obliterating over 40 percent of life on Earth, dinosaurs included.

Other mass extinctions have been related to impact craters. This new evidence has forced us to reformulate the way we think about evolution of life on Earth, our origins, and, ultimately, our survival as a species. Extensive knowledge about cosmic impacts has shaken our beliefs in a somewhat docile, gradual evolution of our planet and its life forms; instead, we now see an evolution punctuated by major cataclysms, which has pushed the "survival of the fittest" idea to its limit. The species with the highest odds of survival were those best adapted not only to their environment but also to abrupt major changes in that environment. The dinosaurs reigned supreme over land, air, and water for over 150 million years, until hell broke loose 65 million years ago. While the reptiles were wiped out of existence, unable to adapt to the harsh new conditions and the disruption of the food chain, small mammals and birds survived. These mammals evolved, and two million years ago the first hominids appeared. We owe our existence partly to that single horrific impact event. Given today's evidence, it is hard to imagine what else would have kept the dinosaurs from still being around. Although we may be justly terrified by the thought of another major impact happening in the future, we should also remember our debt to chance. Collisions destroy and collisions create.

Geology Turns Catastrophic

As an illustration of a true doomsday scenario, we will examine in some detail the fated collision that wiped out the dinosaurs sixty-five million years ago. The story of how geologists, paleontologists, chemists, and physicists came to the conclusion that about 40 percent of life on Earth was indeed killed by a horrendous collision with a comet or asteroid—we are still not quite certain which—is in itself a wonderful illustration of how science works. The hunches, the false clues, the brilliant intuition of some, the severe resistance of others, the accumulation of evidence, the fierce debates—all converged on an eventual acceptance (which is still not unanimous) of a new idea, which, as we have seen, is also very old. Many of the details are told in Walter Alvarez's book *T. Rex and the Crater of Doom*, which is well worth reading.[2] Walter, a geologist, and his father, Luis Alvarez, a Nobel Prize–winning particle physicist, played a key role in unveiling the many clues leading to the inescapable conclusion that the extinction of all dinosaurs coincided with the collision of a 10-kilometer-wide body in the Gulf of Mexico. The record of this "doomsday" collision, and of other collisions, is written in the geological strata of Earth, layer after layer. To quote

Walter Alvarez, "fluids and gases forget, but rocks remember." The challenge was to decipher the message engraved in their memory.

Geologists and paleontologists had long been aware of a fairly sharp boundary separating the Cretaceous and the Tertiary eras 65 million years ago, the so-called KT boundary. However, before the 1960s, very few samples of this boundary were actually available for study. And there was no reason to look for more. Since Charles Lyell published his influential *Principles of Geology* in the early 1830s, geology had been dominated by the idea that changes on Earth occur slowly. Lyell made the curious extrapolation that, since the laws of nature do not change in space and time, which is what makes science possible, processes on Earth should be very gradual, or uniform. This uniformitarian doctrine, as it became known later, assumed that the very slow rate at which things changed on Earth had always been the same. According to this "classical" view, the role of the geologist was to map the different strata, learning about Earth's history as it was recorded on rocks at various heights and depths. This gradualist view gained tremendous impetus during the late 1960s, when plate tectonics became widely accepted as the driving mechanism shaping Earth's crust: nothing seems more gradual than the slow drifting of continent-carrying plates by a few centimeters a year (about the speed at which your nails grow) over eons of time. In this view, the most cataclysmic phenomena Earth could witness were volcanic eruptions, climate changes such as ice ages, and large-scale floods.

Evolutionary biology embraced geological gradualism. Charles Darwin was certainly influenced by Lyell's ideas when he took off on his famous expedition onboard the HMS *Beagle* from 1831 to 1836. When the *Origin of Species* was published in 1859, gradualism was assumed to be a central tenet of evolution: the changes altering a given species were slow, mostly destructive. Although Alfred Russel Wallace proposed similar ideas, it was Darwin who provided most of the evidence and pursued its consequences. Rare beneficial changes helped the lucky species, as it aggressively competed with others for resources and survival. Long necks allowed giraffes to reach the tender higher leaves, spots helped leopards hide better from their prey (and giraffes from their predators), and so on. Natural selection, as this process became known, was a gradual process, leading to the eventual survival of the fittest.

The key ingredient missing in Darwin's theory was genetics. Once genetics and evolution were combined, and DNA was identified in the 1950s, the theory of evolution based on natural selection was understood as a consequence of the faulty transmission of genetic material from one generation to the next. Specific traits of the progeny result from countless variations during the fusion of genetic material from their parents. Most variations are uneventful; puppies all look alike, apart from the location of this or that spot, or the color of an eye. Once in a while, however, a given piece of DNA, a gene, may suffer a *mutation*, a change in its molecular arrangement. This may be caused by different sorts of high-energy radiation (x rays, gamma rays, cosmic-ray particles, natural radioactivity) or by chemical damage. This mutation may be passed on to the individual's offspring. Most mutations cause severe impairment or death. However, on very rare occasions, a mutation may actually benefit the individual's survival. As the mutated fitter organism reproduces, its beneficial traits may pass on to its offspring. For larger animals, which reproduce fairly slowly, it may take millions of years for a species to mutate into an entirely new species, a rate worthy of geological gradualism.

Within this general framework of gradual geological and biological changes, mass extinctions were also believed to be gradual; species don't simply disappear. Several factors, often in combination, may lead to the slow disappearance of a given species, from climatic changes to volcanic activity to imbalances in the predator-prey chain. Thus, it was widely held that dinosaurs disappeared over millions of years, gradually petering out before the KT boundary. The same was believed to be true of the ammonites, marine invertebrates related to the beautiful chambered nautilus of today (see figure 18 in insert).

In an ideal world, one can trace the rise and fall of a given species by examining the frequency with which its fossils are found at different depths; from the bottom up, as a species gradually rises to prominence, its fossil record becomes richer. As the species declines, its fossils become rarer, until they vanish altogether, somewhat like a smooth wave of bones imprinted over vertical layers of rock. An abrupt mass extinction would be characterized by a very sharp decline in the number of fossils of several species. In practice, of course, things are not so simple. Apart from being quite rare, fossils are not laid down regularly, and can easily be destroyed over millions of years. To make things worse, the rarer the fossil, the more gradual the extinction process appears. This is the so-called Signor-Lipps effect; instead of a clearly delineated wave of fossils, one has a flatter wave, whose peak does not stand out much. It was thus reasonable for most paleontologists to argue that the reason why a sharp decline in the number of dinosaur fossils before the KT boundary had not been found was that it wasn't there. However, only a finer combing of the sediments would truly determine the sharpness of the extinction and its location relative to the KT boundary.

In came the catastrophists. In 1971, the paleontologist Dale Russell and the astrophysicist Wallace Tucker bravely suggested that the dinosaurs were killed by a nearby supernova explosion, a huge ejection of matter and radiation marking the end of the nuclear fusion cycle of massive stars.[3] Russell and Tucker conjectured that the explosion could have induced severe climatic changes for which the dinosaurs were not prepared. Other papers followed, analyzing in more detail the consequences of such a violent event for life on Earth. The problem was that astronomers could not find the remnants of a supernova that detonated about sixty-five million years ago in our cosmic neighborhood; the murder weapon was missing. There was also no evidence of the explosion imprinted on the KT boundary. As we have seen, matter ejected during stellar explosions travels across vast cosmic distances, seeding the interstellar medium with heavy elements. Some of it should have been deposited on Earth, like sprinkles on a cake. Luis Alvarez conjectured that a nearby supernova would have left an abnormal amount of plutonium 244—an isotope of plutonium forged during the hellish explosion—over the terrestrial surface. However, no excess of plutonium 244 was found, ruling out a supernova explosion as a viable cause for the KT mass extinction. The gradualists in the scientific community, the vast majority at that point, had the first laugh.

While catastrophist geologists and physicists were busy trying to figure out what would be the telltale substance pointing toward a catastrophic event in the KT boundary, two paleontologists challenged gradualism in their own field. In 1972, Niles Eldredge and Stephen Jay Gould proposed that evolution was far from monotonic, that the fossil record suggests the existence of periods of rapid changes in the number of species, spikes of evolutionary activity, interspersed with long periods of gradual change.[4] Their suggestion of *punctuated equilibrium* was not welcomed by the gradualists; the debate still goes on, although the evidence compiled over the last thirty years indicates that there were indeed periods when new species boomed into existence at an amazingly fast pace. Punctuated equilibrium is a perfect evolutionary compromise between gradualism and catastrophism; an abrupt change in the environment, caused by the impact with an extraterrestrial object or by some other local or, more rarely, global event, would forcefully redefine the notion of fitness. Countless species, previously perfectly well adapted, would find themselves unable to survive under the suddenly imposed new conditions, while others would thrive. The abrupt change in environmental conditions would promote a period of intense evolutionary change, a quick boiling of the genetic soup, until a renewed state of equilibrium was reached, lasting until the next abrupt change.

While these catastrophic ideas were mostly being neglected by the scientific community, Luis and Walter Alvarez and their coworkers were busily rethinking their strategy. After a few false starts, Luis Alvarez came up with a plausible scenario: an extraterrestrial object would be rich in iridium, an element that is very rare at Earth's surface, being mostly concentrated in its molten metallic core. A sufficiently violent impact would vaporize most of the invader and the bedrock around it, launching a huge amount of iridium-rich dust into the atmosphere. Volcanic eruptions do something similar, on a smaller scale: the 1883 eruption of Krakatoa in Indonesia, sprinkled so much dust into the atmosphere that sunsets in London were redder for months, while the global temperature dropped by half a degree Celsius. The dust acts as a reflector of sunlight, cooling the planet underneath. Luis Alvarez reasoned that two effects should follow a huge impact. First, there should be an excess of iridium sprinkled over the KT boundary, brought by the extraterrestrial killer. Second, the dust from the impact would block sunlight so efficiently that the world would turn completely dark and temperatures would plunge below freezing. This is precisely the abrupt widespread environmental change needed to promote global mass extinctions.

The Alvarez team found excess iridium in one known sample of the KT boundary from Italy. Then in another from Denmark. A geologist from Holland, Jan Smit, had independently found the "iridium anomaly" in a KT sample from Spain. In June 1980, the Alvarez team published its paper on the extraterrestrial origin for the KT extinction, using the iridium anomaly as key evidence.[5] A veritable deluge of papers followed. Just during the 1980s, over two thousand papers were published for or against extraterrestrial causes for mass extinctions! The clincher, the discovery of the impact crater, was announced only in 1991. By that time, over one hundred KT sites had been shown to be rich in iridium, confirming the global scope of the disaster. An impact such as this would have left an enormous crater. Where was it? Finding a 65-million-year-old crater is not easy, especially on a planet where almost 70 percent of the surface is covered by water. The odds of having an impact over an area accessible to exploration were small—the opposite of what we would wish of an impact today. Even smaller were the odds of having the crater preserved after millions of years of erosion and seismic activity.

Again, the memory of rocks came into play. The impact, it was later found, happened either very near or just within the waterline of the Yucatán peninsula in the Gulf of Mexico. The enormous violence of the collision caused a tsunami estimated to have grown to at least fifty and possibly a few hundred meters in height. Such a gigantic wave wiped out most of the Gulf coast of the United States and Mexico, as well as the Caribbean, carrying with it all sorts of debris. Working backward, geologists found the sediment layer brought by the giant tsunami right on the KT boundary, at sites in Texas and Haiti. If the tsunami carried sediments to Texas and Haiti, they reasoned, the impact must have occurred in the Gulf of Mexico. Sharing information with geologists from Petróleos Mexicanos (PEMEX, the Mexican oil company), the American geologist Alan Hildebrand and collaborators located the site of impact: the Chicxulub crater, from the Mayan word meaning "tail of the devil," with a diameter of roughly 170 kilometers. It took a 10-kilometer-wide object, traveling at roughly 20 kilometers per second to dig such an enormous hole. It is hard to imagine a more horrific event on a planetary scale.

Rocks have ordered crystalline structures. However, when boiled or vaporized by the tremendous energies of the impact and then rapidly cooled, they fail to rearrange themselves into regular patterns, becoming glassy, or amorphous, in structure. After painstaking search and analysis, glassy materials were identified in samples obtained from several places around ground zero; their ages nicely matched the estimated time of impact at sixty-five million years ago. Some of the crucial work that connected the glassy materials with the KT boundary was done down the hall from my office at Dartmouth by Page Chamberlain and my friend Joel Blum, now at the University of Michigan. In 1993, Joel participated in a debate on the fate of dinosaurs with Chuck Drake, another Dartmouth professor, who nevertheless leaned toward gradualism. The overwhelming evidence for the impact at Yucatán must have made Drake's job very hard. In his defense, however, it should be pointed out that debates such as this, where evidence for and against a given theory is discussed and criticized, are crucial for the well-being of science. I think it is fair to say that Joel won this one.

Dinosaur Doom

The extraterrestrial killer, estimated to have been about 10 kilometers wide, approached Earth at about 20 kilometers per second, crossing most of the atmosphere in less than two seconds. The air in front of it was so severely compressed that it reached temperatures four to five times that of the Sun's surface, creating a blinding flash of light and a sonic boom the likes of which has not been heard since then (see figure 19 in insert). The energy released by the impact was equivalent to an absurd 100 million megatons of TNT, ten thousand times more than all the thermonuclear bombs available at the height of the Cold War detonated together. The consequences of dumping such an enormous amount of energy in so short a time are terrifying; the celestial killer carved into the ocean floor a hole about 40 kilometers deep and 200 kilometers wide, immediately vaporizing all the water and rock it found in its way. Waves of boiling water and rocks rushed into the gigantic hole as 100-meter tsunamis propagated outward. Earthquakes of unheard-of strength, reaching 10 on the Richter scale, shook the ground for hundreds of kilometers, causing the coastline to crumble, while feeding further tsunamis. When we throw stones into a lake, there is always an initial displacement of water, followed by a central peak, which rises higher, the faster you throw the stone or the more massive it is. The impact was so violent that vaporized rock, debris, and water were thrust upward in a plume that went halfway to the Moon, only to fall back again, causing an umbrella of destruction thousands of kilometers wide. Rocks behaved literally like fluids, ground zero marking the site of a giant bull's-eye surrounded by circular undulations that were imprinted on the boiling bedrock.

Any living thing within many hundreds of kilometers from the point of impact was immediately vaporized. Farther out, at a few thousand kilometers, animals saw a blinding flash of light followed by an uncontrollable shaking of the ground, as the seismic waves passed one after another. The ejecta falling back on Earth from space caused the next wave of devastation. They reentered the atmosphere with such fury that the sky was literally set ablaze, igniting continent-size fires that broiled anything in their path. Giant tsunamis completed the immediate devastation. Most parts of the United States and Mexico were completely destroyed in a matter of hours. The heavens brought hell to Earth.

Farther away, in Europe, Africa, South America, and Asia, the immediate effects were less dramatic. Those regions had to wait for the secondary effects of the impact, which were slower, but no less effective, killers. An astonishing amount of dust was lifted into the atmosphere by the impact, as if a million volcanoes had erupted together. As this dust spread through the upper atmosphere blocking the sunlight, Earth turned cold and dark for months. Very cold and dark. So dark that you could not have seen your hand in front of you, and so cold as to reach subfreezing temperatures. The plants and animals unlucky enough to survive the near-instant death from the impact faced agonizing hunger because of the food chain's collapse. Still, there was more to come. As rains washed down the dust, the temperature started to increase. Much of the ground under the impact was made of limestone, a rock rich in calcium carbonate. The heat released by the impact decomposed the carbonate, which combined with oxygen to form carbon dioxide, a greenhouse gas very familiar to us; Earth went from extremely cold to extremely hot, as the excessive carbon dioxide and suspended water vapor trapped the heat from the Sun near the ground, causing an intense greenhouse effect, which lasted for possibly thousands of years. To complete the horror, much of the nitrogen and sulfur found in rocks around the impact area combined with oxygen and hydrogen to form huge amounts of sulfuric and nitric acids, which rained down for decades.

Given the assault from fire to cold and darkness to heat and acid rain, it is remarkable that anything actually survived this true apocalypse, which far surpassed Saint John's most terrifying visions. But several species did, and we are here as their descendants. If the history of life on Earth, in all its myriad forms, can be understood as an experiment in evolutionary genetics orchestrated by natural selection, the emergence of *intelligent* life seems to be the result of a chance occurrence, an odd event that apparently would be extremely hard to duplicate elsewhere. The long reign of the dinosaurs lasted for over 150 million years and showed no signs of weakening until it was terminated almost instantaneously. Their success and longevity as a species make it hard to argue for the necessity of intelligent life at the top of the evolutionary chain; the dinosaurs' limited brains seem to have been quite enough to ensure their dominance. Had it not been for the celestial intruder and the drastic environmental changes that it caused, the dinosaurs would quite possibly still rule, while mammals would be insignificant. It is a humbling thought.

If, indeed, intelligence is not a necessary consequence of evolution, are we alone as thinking beings in this vast cosmos, the odd products of a bizarre evolutionary quirk? Although we will be able to answer this question only by probing for—or being probed by—life elsewhere, it is much easier to argue for extraterrestrial life than for extraterrestrial intelligent life. On the other hand, it is also hard to conceive of a universe with billions of galaxies, each with billions of stars, where any kind of evolutionary quirk can be so rare as to happen only once. I prefer to think of a universe filled with intelligent life, even if spread somewhat sparsely, some of it advanced enough to have found solutions to the many problems that plague our society, such as war, famine, and hatred. Some colleagues argue, quite reasonably, that if intelligent life existed in our galaxy, it would by now have colonized it entirely; the Milky Way, being over ten billion years old and "only" 100,000 light-years across, would be packed full with intergalactic cruisers. Maybe they came a hundred million years ago, saw the dinosaurs, and left in disgust, never to come back. Another popular idea is that we are the results of their seeding, children of extraterrestrial explorers of yonder. If that is true, we all suffer from a bad case of cosmic parental abandonment. Again, I prefer to think that this assumed expansionist need is not an imperative of intelligent life. Perhaps one of the characteristics of the aliens' wisdom is precisely to have found balance within their boundaries, as opposed to the "destroy and expand" policy we seem to pursue so much. How could we possibly presume to understand motive in an advanced alien psyche if we hardly understand our own?

It is tempting to draw parallels between these imagined extraterrestrial beings and the miraculous visions of saints, angels, and gods, popular in many religions across the globe. Some of us *believe* they are "up there" somewhere, wise and distant, a model of what we want or should become. Some persons want to believe so much that they claim to have seen these celestial visitors, hovering over some obscure country road or in a secluded desert—angels or, more often, demons that appear to be as curious about us as we are about them. It is very unfortunate that these alleged visitors never seem to want to establish any sort of serious contact with scientists or political authorities, who could try to establish a true dialogue. In fact, they never seem to leave any concrete proof of their visitations. Their elusiveness makes it hard for most scientists, including this one, to give credence to these sightings. And this is not because we are closed minded. Quite the contrary, as Carl Sagan argued in his book *The Demon-Haunted World*, no one would be more delighted and excited to confirm the existence of extraterrestrial life than the scientists who dedicate their lives to studying the universe and its mysteries.[6] Unfortunately, until that day comes, or until we manage to find it ourselves, intelligent life will remain an odd evolutionary quirk, the end product of a rare celestial collision, and we will have to keep guessing how our cosmic colleagues look and think up there in the heavens.

Apocalypse Now?

Apologies for the sensationalistic rhetoric, but I believe a wake-up call is badly needed. What killed the dinosaurs can kill us too. Yes, scientists have become the new (or at least another class of) prophets of doom. We have followed the early steps of Newton, Halley, and Laplace to confirm the real possibility of a collision with a celestial object. To bring matters closer to our age, let me tell two stories of recent collisions, one here on Earth and the other on Jupiter.

Early in the morning of June 30, 1908, the sky above the basin of the Stony Tunguska River in central Siberia was sliced in half by a blade of light. An object believed to have been an asteroid about 30 meters in diameter, traveling at 20,000 meters per second, exploded at a height of about 8 kilometers above the ground, with an energy equivalent to 15 megatons of TNT, almost a thousand bombs like the one dropped over Hiroshima. A man called Semenov, who was quietly sitting by his porch near a trading station a few kilometers from the blast, was thrown 6 meters away and knocked unconscious. He later recalled that the heat blast was so intense that he thought his shirt was branded onto his chest. The tent over a sleeping family was lifted into the air and dropped down with such incredible force that two people fainted and all suffered bad bruises. The ground shook, and hot winds traveling at high speeds burned and flattened 2,100 square kilometers of forest (see figure 20 in insert). The superposition of two consecutive blasts, the supersonic blast from the object tearing through the atmosphere and the explosion of the asteroid itself, caused the knocked-down trees to assume a curious pattern, resembling the open wings of a butterfly, a dark poetic touch.

At the village of Vanavara, about 60 kilometers away from the blast point, several roofs collapsed, windows were broken, and the incredulous inhabitants saw a huge mushroom-shaped cloud of smoke shoot up into the air, as if a giant conjurer were playing with the elements. Weather stations in Germany detected a pressure wave that circled the globe twice, while passengers in a train 500 kilometers away saw and then heard the event. In western Europe, people observed a peculiar glow in the night sky; even as far away as California, the soot from the blast darkened the skies for weeks. As with the impact that carved Meteor crater in the United States 50,000 years earlier, had this event happened over a densely populated region, millions of people would have died. This is the famous Tunguska event, an eye-opening reminder that these impacts are not entirely things of the remote past. Even more terrifying, it would be extremely difficult to detect such a small celestial intruder with enough lead time to evacuate the target area. A collision like this could happen anytime, anywhere.

The Tunguska event is child's play compared with the devastating collision between Jupiter and the fragmented comet Shoemaker-Levy 9 between July 16 and 22, 1994. This was the best-documented, -photographed, -observed, and -discussed collision in astronomical history; a very rare event that we were fortunate to have the right technology to observe in great detail. Had it happened only a couple of decades back, we would have been confined to distant photos taken by Earth-based telescopes. The Hubble Space Telescope and the *Galileo* spacecraft, which was on its way to Jupiter, provided spectacularly detailed photos, which became instant classics, icons of a new "in your face" astronomy, which blends itself with pop culture through television and, more recently, the Internet: people could watch the collisions in the comfort of their homes, while munching on their snacks and staring at their TV screens (see figures 21 and 22 in insert). It was one of these occasions when terror and excitement get tangled by awe—excitement at being lucky enough to see such a fantastic celestial event, terror at the thought of what would have happened to little Earth had it been the target of the deadly bombardment.

It is fitting that this comet was discovered by the Shoemakers, Gene and Carolyn, and by David Levy—all of them fierce comet and asteroid hunters committed to scanning the skies for possible threats. What Carolyn identified as a "squashed" comet was later confirmed to be twenty-three fragments of one single comet that was ripped apart by Jupiter's enormous gravitational pull, flying together like a string of deadly pearls stretching over hundreds of thousands of kilometers. Like most comets, this one was a dirty snowball, a four-billion-year-old piece of debris left over from the formation of the solar system. Its faraway orbit became destabilized by a rare close encounter or collision with another body, or by passing close to a nearby star or interstellar gas cloud, making it plunge toward the Sun, crossing Jupiter's path. The comet remained in this new, elongated orbit until last century, when it finally got too close to Jupiter; a collision was unavoidable, each fragment crashing into the planet's upper atmosphere at over 200,000 kilometers per hour—over 25,000 megatons of TNT per fragment. Jupiter is eleven times larger than Earth and three hundred times more massive. It does not have a solid surface, although deep down it may have a dense solid core. One can think of it as a giant gas ball, which grows denser nearer its center. What would these cometary fragments do to a celestial body like this?

To make analysis easier, each fragment was assigned a letter in order of impact. Fragment A hit Jupiter's upper atmosphere just before 4 P.M. EDT on July 16. As the fragment blew up in a giant fireball, a plume of ejecta rose to more than 3,000 kilometers above the point of impact. As the debris came crashing down, it left a dark spot roughly a third of Earth's size. And that was just the first collision. This pattern was repeated for each of the fragments observed (a few were lost): initial fireball, rising plume of material, debris crashing down, devastation spreading over large areas. Fragment G, one of the largest, left a bruise roughly the size of Earth.

True, an impact such as this is very rare, especially here on Earth; Jupiter's enormous gravity acts like a kind of cosmic vacuum cleaner for wandering comets and asteroids, decreasing the chances that they will hit us. But the unavoidable conclusion from the Shoemaker-Levy 9 event is that they *can* hit us, with devastating effects. We did not conjecture this event, or reconstruct its consequences from gathered evidence: we saw it! What are we to do?

Searching for Killers

First, we must know what is out there. Collisions can happen with either asteroids or comets, which, as we have seen, are two very different kinds of objects.
Asteroids, found mostly in a region between Mars and Jupiter known as the
asteroid belt, are leftover rocky planetesimals, which could not coalesce into a
larger planet, because of Jupiter's constant gravitational tugging. Comets, also
debris from the formation of the solar system, consist mostly of frozen gases. A
large number of them are found in the Oort cloud, located well outside Neptune's orbit, some fifty thousand times more distant from the Sun than Earth is.
These are the *long-period* comets, which may take anywhere from hundreds of
thousands to millions of years to complete one single orbit around the Sun. We
tend to spot them when they "light up" around Jupiter. The Oort cloud may
have trillions of these frozen objects. In 1951, Gerard Kuiper, working then at
the Yerkes Observatory of the University of Chicago, suggested that another
comet belt existed just beyond Neptune's orbit, which became known as the
Kuiper belt. The objects in this belt orbit the Sun in roughly circular paths
thirty to one hundred times more distant than Earth. Over sixty of these objects
have been found so far, and some estimates put their number in the billions. In
fact, Pluto is considered by many to belong to this belt and to be more like a
huge comet than like a bona fide planet. Clyde Tombaugh, who discovered
Pluto in 1930, calls it "King of the Kuiper belt." These are the *short-period*
comets, with orbits under two hundred years, although some of the Oort cloud
comets can also be captured in orbits passing close to the Sun; it is often quite
difficult to tell which of the two belts a given short-period comet comes from.

In order for comets and asteroids to become a threat, they must somehow be
nudged off their distant orbits. For asteroids, the disturbance could be due to a
collision (or near-collision) with a neighbor or to a passage too close to Jupiter
over several orbits; at every pass, the asteroid receives a gravitational jolt, very
much like a child pumping on a swing. The same way that pumping at the
right time propels the child higher and higher, whenever an asteroid passes sufficiently near Jupiter, it gets pushed to a more unstable orbit until finally it
breaks loose of the asteroid belt. Fortunately for the child, the swing is fastened
to its supporting frame and she won't fly off. For comets in the Oort cloud,
whose huge size makes collisions between neighbors rare, the disturbance is
more likely to be caused by the gravitational pull of a nearby star or of an interstellar gas cloud. In the Kuiper belt, where collisions are still very rare, the pull
from the massive outer planets, especially Neptune, may cause comets to derail.
Once the orbit of a comet or asteroid is sufficiently disturbed, one of two things

will happen: it will get pushed either outward or inward, toward the Sun. The latter case is of most concern to us, since some of these invaders from beyond may settle into Earth-crossing orbits, that is, their new paths around the Sun may intercept ours, increasing the chances of a collision at some point in time. Asteroids that settle into Earth-crossing orbits are known as *Apollo asteroids*, after the first Earth-crossing asteroid, the 8-kilometer-wide Apollo, discovered in 1932, then "lost," and found again in 1980 by chance, when it came within 9 million kilometers of Earth (around thirty times the Earth–Moon distance), pretty close by astronomical standards (see figure 23 in insert).

Now that we know what the threatening celestial bodies are and where they are located, we must get an idea of their number and sizes. For comets, this is an impossible task; there are too many "dirty ice balls" at the Oort cloud and Kuiper belt, and they are too small and too far for direct detection. The typical size of a comet nucleus is a few kilometers, large enough for an extinction-level impact but very hard to spot at distant orbits. For example, comet Swift-Tuttle, a well-known Earth-crosser responsible for the annual Perseid meteor shower, is about 25 kilometers in diameter, more than twice as large as the object that killed the dinosaurs. Another worrisome property of comets is the randomness of their orbits; whereas asteroids rotate around the Sun in the same direction as Earth, comets may rotate in the opposite direction. As we know, hitting something head-on is much worse than being hit from behind; Swift-Tuttle flies by with a speed of 61 kilometers per second; an impact would be equivalent to one billion Hiroshima bombs, a true doomsday event. Fortunately, calculations indicate that the comet is safely out of synch with us; when it visits us again, the nearest approach will occur on August 14, 2126, at a nice long distance of 24 million kilometers. However, when these calculations are run forward far enough, a very close encounter is predicted for 3044, when the comet will pass by at only 1.5 million kilometers, about four times the distance to the Moon. Since cometary orbits may shift because of hot-gas emission from their nuclei, it is hard to be sure about this prediction; very small orbital displacements can make a huge difference after a long time. Moreover, since we know the comet's relative speed to Earth, we see that it has a window of only about 3.5 minutes to hit us, the time for an object traveling at 61 kilometers per second to hit a moving target with Earth's diameter (12,756 kilometers). And 3.5 minutes is a very small fraction out of an orbit longer than a century: the odds for a collision are very small. But how small is small enough for comfort?

For asteroids, the situation is somewhat better, since we know roughly how many and how large. There are three giant asteroids, Ceres, Pallas, and Vesta, with diameters of 913, 580, and 540 kilometers, respectively. Then there are about a dozen or so with diameters over 200 kilometers. We believe that most, if not all, asteroids with diameters of 100 kilometers or more are known and cataloged, as well as about 50 percent of the asteroids larger than 10 kilometers in diameter. In fact, more than 9,000 asteroids have been numbered and named at the time of this writing. Of these, more than 220 are larger than 100 kilometers, and estimates put the numbers of rocks larger than 10 kilometers at over 10,000. As we go down in sizes, the numbers climb fast; there are over 752,000 larger than 1 kilometer, and about 28 million larger than a football field. The asteroid that caused the Tunguska event belonged to this last group of light-weights. Given the enormous distances involved and the relatively small diameter of Earth, the likelihood of a collision may be small. On the other hand, the number of potential strikers is quite large.

Not that the asteroid belt is like a crowded interstate. If we were to build a scale model of the solar system, where the Sun is shrunk into a twelve-inch globe, Earth would be the size of a marble at 100 feet from the globe, Mars a pepper seed at 160 feet, and Jupiter a golf ball at about 550 feet. Now, the mass of all the millions of asteroids put together amounts to about half the Moon's mass, and corresponds to a grain of sand in this scale model. To get an idea of how crowded the asteroid belt is, smash a grain of sand into millions of bits and spread it between the orbits of Mars and Jupiter, located between 160 and 550 feet from the Sun. Each asteroid would be one thousand times smaller than an amoeba; there would be plenty of room for one trillion billion of them, although there are only tens of millions —the asteroid belt is mostly empty space. The images we see in movies or drawings of spaceships dodging swarming asteroids are incorrect, even if exciting. Still, once in a long while collisions between neighbors do happen, and they are launched toward the inner solar system.

Earlier I said that estimates put the number of asteroids larger than 10 kilometers at over ten thousand and those larger than 1 kilometer at almost one million. Of these, Gene Shoemaker and collaborators estimated that at least two thousand are Earth-crossers, that is, potential hitters. In the early 1970s, the Shoemakers, working for the U.S. Geological Survey, and Eleanor Helin, from the Jet Propulsion Laboratory in Pasadena, led two teams of asteroid-hunting astronomers. Their strategy was fairly simple: one can find asteroids by photographing the night sky with a telescope at regular intervals; since they are rocky, they reflect sunlight just like planets, showing up as dots of light moving against the background stars. An asteroid near Earth will appear to move faster against the stars than one farther away. If you have a small telescope and have ever tried looking at Jupiter and its moons, you can imagine how hard it is to spot rocks only a few kilometers wide and many millions of miles away. In

1989, two members of a spin-off survey program called PACS (Palomar Asteroid Comet Survey) discovered an asteroid about half a kilometer wide a few hours *after* it brushed by at a distance of only about 600,000 kilometers, just 1.5 times the distance to the Moon. The asteroid, now named Asclepius, would have caused severe damage to the whole planet. It will probably hit us some day, like countless other Earth-crossers.

Shoemaker's and Helin's pioneering efforts were far from enough; there are just too many Earth-crossers out there. In the early 1980s, Robert McMillan and Tom Gehrels, from the University of Arizona, created the Spacewatch Project: a dedicated 36-inch telescope at Kitt Peak, Arizona, armed with extremely sensitive light detectors known as charged-coupled devices (the CCDs you see in video cameras), capable of finding six hundred asteroids in one good night.[7] Another program flourished for a while in Australia, but had its funds cut in 1996. In fact, of the several original programs started in the seventies and eighties, only Spacewatch is still operational. Even with its advanced automated search technology, it would be hundreds of years before it could track most of the Earth-crossers, a luxury we cannot afford.

In the early nineties, a larger international effort, known as Spaceguard Survey, was proposed, following a request from the U.S. Congress to NASA to examine the impact question. The idea was to have a network of dedicated telescopes, which jointly could track over a hundred thousand main-belt asteroids per month, as well as rarer Earth-crossing comets, way before what we can do with current technology (but no more than about a year or two before impact). With six 2-meter telescopes, this warning system would cost about $50 million to build and $15 million a year to run, a modest sum compared with what is spent in countless research projects worldwide. Under the original plan, Spaceguard could reduce to a couple of decades what would instead take centuries with current search tools. In the meantime, an organization called Spaceguard Foundation, with headquarters in Rome, Italy, was established by the international astronomical community to integrate observations and data analysis from observatories around the world. The goals set by Spaceguard are ambitious: to have 90 percent of all near-Earth asteroids (NEAs) larger than one kilometer, tracked by 2010. This requires an eightfold increase over current search rates.

Although Spaceguard has not materialized as originally proposed, there are currently three NEA searching programs, apart from the ongoing Spacewatch, which have been quite successful: LINEAR, from the Massachusetts Institute of Technology Lincoln Laboratory and the U.S. Air Force; NEAT (Near Earth Asteroid Tracking) carried out by the Jet Propulsion Laboratory and the U.S. Air Force and headed by Eleanor Helin; and LONEOS, from the Lowell

Observatory in Massachusetts. Together, these programs doubled the number of known NEAs since 1998; as of August 2000, we knew of about 1,050 NEAs, of which 400 are large ones, that is, wider than 1 kilometer. Recently, NEAT astronomers have revised the estimate of the number of NEAs larger than one kilometer, lowering the old figure of about 2,000 objects to under 1,000, a 50 percent cut in potential hitters.[8] That being the case, we are roughly halfway toward the Spaceguard goal, not all that bad. One may legitimately question, however, whether this goal is truly foolproof.

Statistics are often quoted both to raise and to lower the risks of a collision. Events such as the one in Tunguska or the one that carved Meteor crater are estimated to happen once every century. Objects between 100 meters and 1 kilometer wide may hit once every 100,000 years. Objects larger than 1 kilometer, the ones capable of global damage, may hit once every 10 million years. A first look at these numbers tells you not to worry: the odds for collisions with large objects are very small. However, these numbers are obtained by averaging over known impacts on Earth and on the Moon, and from our current inventory of Earth crossing objects. *They do not represent a prediction, but an expectation.* And, as we know, even when the odds are small, people win lotteries. It would be quite foolish to rest the future of civilization (at least of countless lives) on the feeble assurance of small odds. It is a matter not of whether a serious collision will happen but of when.

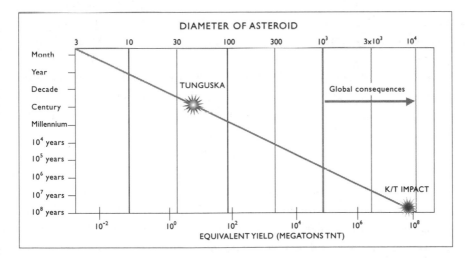

FIGURE 7: *Diagram of asteroid impact energy* (bottom) *and frequency* (vertical) *versus size of the object* (top). *A Tunguska-class event, at about 30–50 meters, happens on the average once every century, with an energy of tens of megatons. The thicker vertical line marks impacts with global consequences.*

Fighting Doom in Outer Space

On May 18, 1996, the headline of the *Boston Globe* read, "Thank heavens, huge asteroid is only passing by." The asteroid, measuring 500 meters across, brushed by at 446,000 kilometers from Earth, a whiff more than the distance to the Moon. It was the largest object passing so close ever recorded, and it was discovered only five days before its nearest point of contact. Imagine what Martin Luther or Increase Mather, whom we met in chapter 2, would have made of this: "God has sent us a sign! Sinners, repent before the millennium comes, or face his wrath and spend eternity in the pits of hell." What was first confined to ancient religious texts, prophecies of fiery rocks falling from the skies and bringing widespread destruction, has become a legitimate branch of astronomy. During the last two decades of the last century, Earth scientists confirmed that a huge impact killed the dinosaurs, Congress urged NASA to seriously gauge the impact threat, and cosmic collisions (like that between Shoemaker-Levy 9 and Jupiter) and near-misses were actually seen and recorded in great detail. In many people's minds, old apocalyptic fears were awakened, and by none other than the scientists themselves; yes, astronomers have become prophets of doom. But a true prophet would not just foretell doomsday but also offer a path to salvation: according to astronomers, salvation could come from destroying or deflecting an incoming celestial killer in outer space.

Hollywood did not waste any time. If real asteroids were not about to hit us, at least a barrage of impact-disaster movies was. In 1997, the American television network NBC produced the two-part *Asteroid*, where Earth braces for impact with two asteroids, a smaller one that hits Kansas City, Missouri, causing damage over a 200-mile radius, and the huge Eros, a 6-kilometer-wide monster. A secret laser gun (no doubt inspired by the flawed Strategic Defense Initiative project) was to blast the asteroid to pieces. Thousands of fragments end up raining down all over Earth, causing terrible, widespread damage. The largest fragment, 80 meters across (more than twice as large as the Tunguska stone) falls right on Dallas, wiping out the whole city. Blasting asteroids is, in general, a very bad strategy.

Although the movie had several flaws (one being that although over 70 percent of Earth's surface is covered by water, the large fragments tend to fall over American cities), it does alert the population to two crucial facts: that collisions with Earth can happen, and that we can do something to protect ourselves. To a certain extent, disaster movies serve as a sort of unconscious rehearsal for confronting some of our hidden fears; we see it happening in a movie, knowing all the time it can also happen in reality, an element of plausibility lacking in horror movies. As such, disaster movies fabricate an "almost reality," awesome in its tragic power but still everyday enough that audiences can identify with it. Planes crash, skyscrapers burn, earthquakes shake the land, and volcanoes erupt in the real world and in the movies. And now the newcomer to the list of plausible accidents is "death by impact," changing the destruction scale from local to global. Fiction and science mix with great dramatic impact. Only in the Bible can we also find such violence associated with cosmic collisions.

In the summer of 1998, America was hit by two huge blockbuster movies exploring the consequences of global-killing collisions, *Deep Impact* and *Armageddon*. In the more interesting *Deep Impact*, featuring Morgan Freeman as an African American president, a 7-mile-wide comet was due to hit within two years. Again, a secretly designed spacecraft—called, appropriately enough, *The Messiah*—is deployed to meet and destroy the comet. Its mission is to land on the comet, dig a hole into its core, implant nuclear explosives there, and detonate them remotely, in the hope of blasting the comet into many small and harmless fragments. Unfortunately, the blast succeeds only in splitting the comet into two pieces, both still hurtling toward Earth. (A good illustration of conservation of linear momentum.) The smaller one, measuring 1.5 miles, hits the Atlantic, devastating all eastern North American coastal cities and presumably a lot of Africa and Europe as well. Since a mission named *Messiah* cannot possibly fail, in a desperate last attempt its crew decides to collide head-on with the larger 6-mile-wide piece, detonating its last four nuclear bombs: martyrs dying to save civilization.

Then came *Armageddon*, a more typical action-packed crowd pleaser, where ordinary people, Bruce Willis and his crew of oil diggers, become heroes under extraordinary circumstances. This time it was an absurdly large asteroid "the size of Texas" (no such asteroid would have been around without being noticed) to hit in only eighteen days! Willis and his crew are hoisted onto the hellish-looking metallic monster asteroid ("Dr. Seuss's worst nightmare," cries a crew member), complete with dark spiraling peaks, to dig a hole and plant nuclear explosives inside. Armageddon —where good and evil have their final battle— is the asteroid itself, a flying piece of hell, confronted by the crew of a spaceship that resembles an angel; the religious symbolism is quite obvious. Again, the mission is accomplished at the last minute by an act of heroic martyrdom.

We can now go back to a more sober analysis of what could be done if one of the NEA searches spotted an object traveling toward us. With present-day or near-future tracking technology, there are three possible scenarios, which call for different approaches. One possibility would be to find a fairly "small" asteroid, under 500 meters wide, due to impact within a few months. Being relatively small and faint, an object like this could pop up on very short notice, as we have learned during the past few years. Unless we had an intercepting mission completely ready, our only option would be to track its orbit as precisely as possible, find the impact date and site, and evacuate if necessary. A second possibility is similar to the first one, but with much more serious consequences; to spot a long-period comet with one or two years' notice. The main difference here is size: comets are on the average a few kilometers wide, bringing the impact into the global-cataclysm category. Without an intercepting mission ready to go, we could do little more than stock up food, seeds, and provisions and build underground shelters with at least one or two years of functionality. These measures would probably guarantee the survival of our species. It would be quite difficult to decide who would go into the shelters, a problem brought up in the movie *Deep Impact*.

The last possibility, by far the most favorable, is to detect a large asteroid with a few decades notice, as is Safeguard's goal. Here, even if we did not have any defenses set up, we would have ample time to prepare. Taken together, these three scenarios indicate the need to intensify our search capabilities in order to have as much lead time as possible, certainly no less than a few decades. This translates into building more telescopes dedicated to searching for near-Earth objects (asteroids and comets that pose a threat), preferably with larger mirrors (2 meters plus) so as to catch fainter objects, be they smaller or just farther away. As for developing the technology to intercept and somehow get rid of the celestial killer, here things get more complicated, as science becomes directly enmeshed with military strategy.

Prepare or Else!

Ideally, we would have such great search capabilities that most threatening objects would be spotted with a few decades or more to spare, as in case three above. This is the joint goal of all present search programs and others planned for the near future. The crucial question, however, remains: Is this "react only if under threat" approach enough? Spaceguard's original document clearly states that a survey based on several 2-meter telescopes would have a success rate of about 75 percent; that is, we could still miss one out of every four approaching objects. Even if this were an acceptable margin of risk for asteroids and short-term comets, recall that long-period comets are found only as they get close to Jupiter, giving us at best a few years' advance notice. I would argue that our present search programs are still wanting; the stakes are just too high. If we were to take the search for potential cosmic hitters to a new level of much needed precision, we could perhaps follow Freeman Dyson's suggestion and develop space-borne telescopes combining very wide fields and powerful optical fiber technology, the heir to the wonderful Hubble Space Telescope. These tools would, in principle, be capable of detecting very faint and distant objects, increasing our lead time from decades to centuries.[9] This very prudent strategy would also greatly benefit many branches of astronomy, from planetary to extragalactic. Intercepting missions cost much more than space-borne telescopes and, with luck and reliable telescopes, could be avoided for centuries or even millennia.

A serious dilemma presents itself at this point. If there is no plan to fund such wide-field orbiting telescopes, should there be a contingency plan for a rendezvous with an incoming celestial killer? Such a plan would be tied up with national defense and security, being coordinated jointly by the U.S. Air Force and NASA. As I write these lines, President Bush had just decided to give the go-ahead to the National Missile Defense (NMD) system, an extremely costly program (estimates place it in the $60–100 billion range), which is supposed to intercept incoming missiles from rogue nations, such as Iraq, Libya, or North Korea. The scientific community overwhelmingly opposes the project, not only idealistically but technically; it is easy to argue that such defense systems are bound to fail, since they will not be able to differentiate between bomb-carrying missiles and increasingly sophisticated decoys, which would be released with the attacking missiles. Experts claim that decoys and other coun termeasures would quickly saturate the defense system. Moreover, if a terrorist nation really wanted to attack the United States, it could do so with chemical and biological weapons, which could be delivered in Federal Express packages, as opposed to missiles. (Recent events have painfully shown that terrorism does not need missile technology to be effective.) The political consequences of such a unilateral development are very serious, as nations like China have promised to respond to the creation of an NMD system by increasing their own nuclear arsenals: what should defend creates more tension and possible grounds for future conflict.

In the Hollywood movies I discussed, the intercepting missions were all improvised spin-offs from American defense projects, not unlike the NMD. Instead of building a missile defense system unilaterally (or almost), perhaps we should create a truly international collaboration for the detection and interception of rogue asteroids and comets—a League for Planetary Defense, dedicated to the construction of large telescopes and an array of surveying and intercepting missions, designed for different situations. If the threat is global, we should react globally, as with more immediate global threats like pollution. Unfortunately, we tend to act when the threat is all too imminent. This may work sometimes, but with a celestial killer, it won't. The probability of a serious impact is certainly very small, and nobody should take to the streets screaming "The End is Near!" on the basis of this or any other book on the subject. However, the odds are there, as the Shoemaker-Levy 9 episode gruesomely reminded us. I take it as a warning sign, a scientific equivalent of the religious "Repent or else," which could go as "Prepare or else." We have the chance to do something, and we should.

Suppose astronomers find an incoming asteroid or comet. What should we do? The first response is usually, "Well, we fly out there and blast the bugger into smithereens with our nukes." Not a good idea, even though with current technology it might be our only choice. Explosions transfer a huge amount of energy but not in a specific direction. If several bombs explode near or inside an asteroid or comet, managing to break it into pieces, something that depends crucially on the composition of the object, the odds are quite high that some or most of the pieces would continue to move in the same direction, raining down on Earth with devastating effects. Without enough detailed knowledge of the celestial object's composition and geological structure for us to be able to predict how it would fragment, explosions are quite a gamble. For example, some asteroids are put together as "rubble piles," somewhat loose and porous assemblages of smaller parts. These asteroids are naturally more resilient to impact, and may absorb the energy of an explosion much better than hard, rocky ones. However, nuclear detonations or just a head-on collision with a large spacecraft may be able to deflect smaller objects, those posing only local threats. Unless they are zooming toward a densely populated area, it is an open question whether any action in space is justified.

The best approach is to transfer momentum to the object, deflecting it as far from Earth as possible: a small deflection at large distances causes a large miss, as any billiards player knows. This is because momentum transfer involves direction, and can push the object away from a colliding trajectory with Earth. Just as auxiliary rockets are used to position a spacecraft into different orbits, we can envisage different mechanisms to gently nudge the celestial object from its killing path. One of them involves implanting rockets on the surface of the object, transforming it into a giant spacecraft, which could then be guided into a safe trajectory (see figure 24 in insert). Such a project will doubtless face tremendous technical challenges. For example, what powerful rockets would these be and what fuel would they use? The first answer is, again, nuclear power. However, the development of such advanced nuclear technology goes against the current trend toward nuclear disarmament. Also, placing nuclear power in space has many potential detractors, because accidents could have serious environmental consequences. The first of these problems could possibly be resolved with a truly international collaboration, but the second one remains. (Unless the asteroid is rich in uranium, which could then be locally mined and processed into nuclear fuel.)

Another idea is to use the object's own materials to push it away. A digging device could be developed that would pump matter from the object's surface and expel it with large velocities in a given direction, somewhat like a huge hose. How to power such machines, known as mass drivers, without using nuclear fuel might be a difficult problem. The cleanest solution is, of course, solar energy. However, this requires installing huge collectors on the surface of the object, and their efficacy would be restricted to interceptions sufficiently close to the Sun that enough power could be collected. To make things more complicated, both asteroids and comets tumble quite erratically as they move along, making a steady collection of power for even a few hours impossible. Furthermore, the surfaces of these objects, especially comets approaching the inner solar system, are notoriously hostile, with violent jets of vaporized matter popping all over the place, "Dr. Seuss's worst nightmare." It might be quite hard to install and maintain such facilities for a sufficiently long time.

Problems and challenges notwithstanding, we must plan ahead. I have no doubt that if sufficient resources and brainpower are put together, solutions to most problems will be found, just as they were for the manufacture of the atom bomb by scientists of the Manhattan Project. An expanded fleet of space-borne wide-field telescopes and survey spacecraft would give us plenty of time to explore the incoming object's composition and decide on the best course of action. The recent success of NASA's NEAR (Near Earth Asteroid Rendezvous) mission, in which a spacecraft operated from Earth orbited and landed on an asteroid for the first time, 433 Eros (not the one in the movie *Asteroid*), gives us plenty of confidence that we have at least the preliminary technology needed for interception (see figure 25 in insert). Different contingency plans should be designed for action on different objects.

Since the pioneering work of Newton, Halley, and Laplace, we have learned not only to foresee the dangers associated with cosmic collisions but also that we can do something about them. As I have tried to show, during the last few decades we have amassed convincing evidence that such collisions occur all over the solar system, Earth included. In the words of Increase Mather's sermon title, they are "heaven's alarm to the world." Should we just put this information behind us and pray to God that no such horrendous events ever occur?

Now, we should take a step back from our self-centered position and recall Newton's wonderfully liberating vision of cosmic renewal and destruction. Comets, he argued, weave the universe together, feeding spent worlds with their generous vapors. We have seen how life on Earth has been stirred into spurts of amazingly fast-paced growth by such cosmic collisions and their after-effects. The end of an era marks the beginning of another, creation following destruction, in a dance largely orchestrated by celestial dynamics. We have learned to love its beauty, inventing time to follow its cadence in ways we can understand and quantify. But if we take the next step, peering out of the solar system, we find that our beloved Sun is just an ordinary star, fated to also become one of the spent worlds Newton hoped comets could resuscitate. Alas, it is not so. As we move outside our immediate cosmic neighborhood, we should invoke another vision, that of Kant, in which stars and their courts of planets are constantly being created and destroyed in the vastness of space, like the mythic bird Phoenix. So, we take off to the stars now, where one day we must find our new homes, if we are smart enough to survive the many forthcoming cosmic collisions in the ensuing five billion years before the Sun evolves into a new stage and life on Earth collapses forever.

PART III

The Rise and Fall of Stars

CHAPTER 5

Fire in the Sky

When thou hast risen they live,

When thou settest they die;

For thou art length of life thyself,

Men live through thee.

—CHANT TO THE SUN FROM THE REIGN OF

AKHENATON IN ANCIENT EGYPT (CA. 1370 B.C.E.)

We all gathered on the ship's upper deck, carrying our cameras and protective lenses; it was August 11, 1999, total solar eclipse day, the last of the millennium. The aptly named *Stella Solaris* (Solar Star) was flanked by three other large ships, which together brought thousands of people from all over Europe and the United States to witness what is surely one of the most awesome natural events. Many were "eclipse groupies," addicts who travel across the globe hunting for total eclipses, trying to quench their thirst for the unfathomable. My wife, Kari, and I were "eclipse virgins," for our knowledge through books and photos was much bigger than our experience of the "real thing." It was our good fortune that I was chosen to lead a group of Dartmouth alumni on this cruise, while Kari, an alumna herself and very skilled at social dynamics, worked to make sure everyone was happy. Our most recent encounter with an eclipse had been only a partial one, visible from New Hampshire during the spring of 1994. "Wait until you see this," was what we heard over and over again from several "groupies," as the ship sailed the Black Sea waters in search of the optimal position, clear of clouds and well within the umbra, the cone-shaped shadow cast by the Moon as it blocks the sunlight.

I had seen a total eclipse before, when I was seven, holding my nanny's hand and slurping on a Popsicle given to me by "Seu" Alexandre, the corner Popsicle vendor with a quixotic countenance. We were planted right under my building, a block away from Copacabana Beach, surrounded by a huge crowd of fellow "eclipse virgins." (The eclipse was actually partial in Rio. To my seven-year-old imagination, however, it was as grand as the real thing, and registered as such.) I remember mostly the reverent silence that befell every one of us as the Moon blocked more and more of the Sun, as if our voices had been fed by the waning sunlight: even though it was midmorning, the birds stopped chirping, and the traffic in the busy streets froze as if by magic (a true miracle in Rio, repeated only during World Cup games). High above, a dark disk surrounded by a dif-fuse yellowish light looked to me like the eye of a very angry and powerful God, not too happy with what he saw below. "Is the Sun dying?" I asked, echo-ing age-old fears that inspired countless myths. After a couple of minutes of agony, a thin and gracefully curved line of bright light appeared from behind the dark disk, forcing me to look down, a big smile of relief stamped across my face. Light won after all; God gave us another chance.

I was thinking of these vague but powerful memories as we waited for the "first encounter," when the disk of the Moon just touches the golden solar disk, initiating the eclipse. Would I be as scared and awed as when I was seven? Or would all these years of science have trained my mind to look at nature with the steadier eye of reason? We were told to watch for the beautiful "diamond ring," which happens just before totality (and just after) when the dark lunar disk is surrounded by a ring of light spiked by an incredibly intense beam, the "dia-mond" (see figure 26 in insert). We should also strain our eyes to find solar prominences, bursts of red flaming gas that shoot up from the solar surface, which are hard to see unless the glare from the solar disk is blocked. Lastly, as daylight fades away during totality, we should look for stars and planets, espe-cially Mercury and Venus, and for the translucent corona, the tenuous outer layers of the solar atmosphere. All this in one minute and twenty-two seconds, the time of totality for this eclipse.

If the relatively short time for totality was not ideal, at least we were lucky with the weather, which was perfect, and with the time of the eclipse, which happened around noon, when the Sun was high up in the sky. I was somewhat stressed out by the many things I was supposed to do and look for, which also included taking pictures of the eclipse. As the Moon proceeded to block over half of the solar disk, the daylight dimmed quite considerably, tinging every-thing with a vague graininess, as if reality were about to dissolve right in front of our eyes. A touch of angst invaded me and, from what I could gather, most people around me; the exceptions were those experienced eclipse-goers who, juggling telescopes, zoom lenses, and many different cameras, were too busy to wax existential. At this point, the Moon had mostly covered the Sun, and we prepared for the show.

A diamond ring exploded around the dark disk of the Moon, even more spectacular than I expected, its incredibly intense light etching its patterns onto our memories forever. I looked around; the sky had turned into a majestic metallic blue, and two dots of light were barely visible around the darkened Sun: Venus and elusive Mercury. We had a 360-degree view of the horizon, which turned into a pink band, as if the sunset were coming from all points in space at once. The breathtakingly beautiful combination of colors and darkness brought tears to my eyes, and I was invaded by a completely irrational reverence for what was happening; there was no room for the steady eye of reason now. Kari, standing by my side, was pretty much in an ecstatic trance. Before I followed her, I managed to take a couple of pictures. And then it was time to look up.

There was the eye of God again, staring at us down below. To the Persians, the Sun was the eye of Ormuzd; to the Hindus, the eye of Varuna. To the ancient Greeks, it was the eye of Zeus; to the early Teutons, the eye of Wodan (or Odin). I was invaded by a soft sense of terror, at once primal and sublime, as the silver-blue darkness engulfed us all, a light from the world beyond, a light from death: here I was, thirty-three years later, staring once more at eternity. A total solar eclipse is a highly subjective experience, each person's reaction no doubt a mirror of what he or she carries inside. Filled with awe, I could easily understand why eclipses were viewed with such horror in so many past cultures; the event's primal energy epitomizes the archetypal struggle between good and evil, light and darkness. To some, a dragon was eating the Sun; to others, a serpent. But to all, the event symbolized our indebtedness to the Sun, the life-giving god. More than that, it symbolized our complete dependence on the Sun, our fear that if it goes, we go with it. The threat of a darkened Sun mirrored the threat of the end, a symbol readily adopted by the apocalyptic narratives we analyzed earlier on. (Recall Revelation 6:12–14, with the darkened Sun, and its depiction by Luca Signorelli shown in figure 3 in the insert.)

As I hope to have made clear in the first two parts of this book, science, in its thematic development, incorporates trends from popular beliefs and religion. For example, the end of Earth, prophesied by several apocalyptic texts as being promoted by cosmic cataclysms ordered by angry gods, found its expression in a scientific analysis of the possibilities of future collisions with asteroids and comets. Of course, this is not a one-way road, and scientific trends are also incorporated by religious discourse; one recent example is the tragic end met by members of the Heaven's Gate sect. In the present chapter, we will keep following this trend, investigating how worship of the Sun in so many cultures, and the preoccupation with its stability, filtered from popular beliefs into scientific discourse. The question of how and for how long a star shines is as pressing for a modern-day astrophysicist as it was for the Lord Inca.

Sun Worshipers

The Sun must have been adored since men first realized that their survival chances were enhanced if they gathered into groups. There is such a rich profusion of religious rituals, myths, and legends dedicated to the Sun throughout recorded history that it is safe to say that every religion, past or present, has had a role for the Sun in its structure. Among its many human or partly human incarnations, the Sun was Apollo for the Greeks, Thor for the Vikings, Horus, Osiris, or at times Amon-Re or Atum for the Egyptians, the Sun goddess Amaterasu for the Japanese, while, in early Hinduism it was represented by Suryah, Savitar, and Vishnu.[1] It would be a futile task to present here a comprehensive overview of Sun worshiping through the ages, although the topic is truly fascinating. Instead, we will examine three illustrations of Sun worshiping, in Egypt, Japan, and South America, exploring how they express the complex relationship between ourselves, the vast cosmos we inhabit, and the passage of time—or, more specifically, how we cope with our perplexity at being able to ponder the eternal, only to succumb to death.

For the Egyptians, there were two gates in the sky, one to the east, where the Sun god came out everyday, and one to the west, where he retired for his descent into the underworld. This diurnal motion of the Sun was viewed with a great sense of mystery, the disappearance of the Sun god marking his defeat to the demons of the underworld, whereas his reappearance at dawn marked his victorious return. This cycle of death and resurrection was mirrored by the human soul, for it too had to battle the demons of the underworld before it could ascend triumphantly back to life. During the first and second dynasties (3100–2600 B.C.E.) the pharaoh was a god—specifically, the incarnation of Horus, the all-powerful, life-giving Sun. This identification of the pharaoh with an omnipotent deity certainly contributed to the centralization of power in ancient Egypt. Echoes of this doctrine can be found in many other cultures— for example, in Japan, where the emperor is a direct descendant of Amaterasu, the Sun goddess, and with French absolutism, which reached a climax during the reign of Louis XIV, the "Sun King," the longest in European history, lasting seventy-two years! It is no small measure of man's arrogance to justify his political power through the omnipotence of the Sun.

Back to Egypt, during the Old Kingdom (2600–2200 B.C.E., also known as the Pyramid Age), the most adored god was Re, the chief Sun god, the "father" of the pharaoh. Egyptian worship was firmly centered on the Sun, even though the related deity changed in time and space. Osiris ("the one who sees clear") was the most beloved of the solar deities, associated both with the setting Sun, when he descended into the underworld, and with the Sun above. He would thus connect the world above with the world below, life and death, the central tenet of Egyptian religion mirroring the daily motion of the Sun.

This adoration of the Sun reached its peak during the reign of the pharaoh Akhenaton (akhen, pious; aton, solar globe), who courageously declared all deities false, imposing a monotheistic cult of the Sun all over Egypt. The modern American composer Philip Glass wrote an opera inspired by this unique episode in Egyptian history, which no doubt sent monotheistic waves across the Middle East. But it was to be very short-lived. Akhenaton's successor, his son-in-law Tutankhaten, changed his name to Tutankhamon, reinstating the cult of Amon-Re and the plethora of other Egyptian deities. In this reinstated polytheistic cult, the Sun continued to reign supreme, until Christianity overthrew its "pagan" adoration.

Perhaps the most poetic of all expressions of reverence to the Sun can be found in the Shinto ("The Way of the Gods") religious tradition of Japan. According to one of the versions of the Shinto myth, the Kojiki, or Chronicle of Ancient Events, the Japanese islands were the creation of the male and female primal gods, Izanagi and Izanami. After the primordial chaos was separated into heaven and ocean, Izanagi descended on the Floating Bridge of Heaven (possibly a rainbow) and stirred the waters together into an island. Izanami then joined him, and bore the eight islands of Japan and several other deities, until she died giving birth and went to the underworld. Terribly distraught, Izanagi followed her, hoping to bring his companion back. But he was too late; she was already decomposing and felt shamed to be seen in this state by her former partner. Embittered, she sent horrible monsters to chase Izanagi out of the underworld. After many adventures, he escaped and went to bathe in the ocean, hoping to cleanse himself from the evils of the underworld. From the washed filth of Izanagi's left eye was born the most revered of all Japanese deities, Amaterasu, the Sun goddess. Out of his right eye came Tsuki-yomi, the Moon god, who proceeded to join Amaterasu in heaven.

Amaterasu sent her grandson Ni-ni-gi to rule over the islands for her, commanding him in words most Japanese children learn in school: "This Luxuriant-Reed-Plain-Land-of-Fresh-Rice-Ears shall be the land which thou shalt rule."[2] Later, Ni-ni-gi's great-grandson, Jimmu Tenno, the first human emperor, took over the province of Yamato, on the central island of Japan, and set up the empire's capital there in 660 C.E. Thus, the Shinto myth establishes the holiness both of the Japanese islands, a creation of the gods, and of the imperial family, whose members are direct descendants of the Sun goddess, Amaterasu. In the Shinto tradition, the Sun is worshiped as the essence of all that is holy. The most revered shrine in Japan, the Grand Imperial Shrine at Ise, is dedicated to Amaterasu. Within the inner shrine of the large temple, there hangs a beautiful mirror, the most precious of all "divine Imperial regalia."

It may seem strange that a mirror occupies such a high place in Japanese tradition. However, its symbolism, as explained in a myth of great poetic beauty, expresses the hope that the Sun will always shine and that darkness will never overcome light. Amaterasu's brother, the storm god Susa-no-wo, was causing major havoc below, ruining rice fields and polluting the waters. The final straw came when Susa-no-wo destroyed the roof of Amaterasu's weaving hall with a thunderbolt and threw a "horse from heaven" through it. Amaterasu's ladies-in-waiting were so terrified that they all died of fright. In disgust, Amaterasu retired to a cave in heaven and shut the entrance behind her. With the Sun hidden away, the world quickly sank into darkness, a mythical re-creation of a total eclipse. And with darkness came decadence, as evil spirits could now run free across the world and cause all sorts of mischief. The "eight million gods" were so worried that they devised a plan to get Amaterasu out of the cave. They brought a sacred sakaki tree to its entrance, and hung many artifacts on it, including an eight-foot-long mirror. Then, Uzume, the phallic goddess, started to dance to the singing of the deities, who laughed and screamed with premeditated joy. The noise made Amaterasu very curious: she stepped out of the cave, expressing her surprise at all this joy and laughter in her absence. "Well," answered Uzume, directing the mirror toward Amaterasu, "we are celebrating the appearance of another beauty to rival yours." As Amaterasu glanced at the mirror, hardly believing her eyes, another god ran behind her and stretched a rope, the *shimenawa*, across the cave's entrance to stop her from going back in. The land below shone bright, and all gods rejoiced in Amaterasu's return. As the mythology expert Joseph Campbell has remarked, if the Christian cross is the "symbol of the mythological passage into death, the *shimenawa* is the simplest sign of the resurrection."[3] Both represent a boundary between two worlds—the world of the living and the world of the dead, the world of light and the world of darkness.

For our final cross-cultural foray into Sun worshiping, we go to South America, specifically to Peru ca. 1100 C.E., where the Incas had established a Sun-worshiping empire as grandiose as the Egyptian or the Japanese. Like the early pharaohs, the Lord Inca was a direct descendant of the Sun god, the Supreme Creator, whom they called Inti (Light). Everything under the sky—the land, the people, the gold—belonged to the Lord Inca, a de facto theocracy. The ruler personified the Sun on Earth (see figure 27 in insert)., bringing his divine presence to the everyday lives of its subjects.

Every aspect of life for the Incas revolved around Sun worshiping; villages were built with an unobstructed view of the east so that its inhabitants could join in worship every sunrise. Of all the many deities of the Incas, only the Sun had a temple in every city. The magnificence of these temples is legendary, perhaps not matched by those of any other nation on Earth (see figure 27 in insert). Gold was extremely special to the Incas, since it was believed to hold a mysterious relation with the Sun; the nuggets found in the mountains were thought to be tears from the Sun god himself. In order to express their reverence, the Incas decorated their temples with enormous amounts of gold. In Cuzco, their most sacred city, founded by the legendary first Lord Inca, Manco Capac, ca. 1100, there stood the Palace of Gold, as the Great Temple of the Sun was known. The whole building, an enormous structure, was surrounded by a thick frieze of gold six inches wide. The same friezes adorned each of the internal rooms. A huge golden disk, with the engraved face of the Sun god and studded with precious stones, adorned the main altar-facing doors opening to the east. At certain times of the year, the disk reflected the sunlight so powerfully that it appeared the Sun god himself was visiting his temple.

Like the Druids, who, as we saw in chapter 1, were also Sun worshipers, the Inca priesthood performed human sacrifices. The blood of the victim, preferably that of a young virgin, was smeared over mountaintop sacrificial stones so as to glimmer in all its gory redness during sunrise. The rituals made sure the Sun god would be sufficiently pleased with his subjects; otherwise, the world would plunge into life-destroying darkness. This darkness finally came in 1535, with the murderous plundering by the Spanish armies of Francisco Pizarro, and the several revolts and disputes that followed. A mercenary soldier apparently stole the golden disk, the most revered object in the Palace of Gold, only to subsequently lose it during a drunken bout of gambling.

To a large extent, most of the Sun-worshiping religions were (and are) concerned with one central theme—that the preservation of life depends on the stability of the Sun, its cyclic motions, and the light and warmth it generates. Worship of the Sun was an expression of the people's indebtedness to its life-giving powers and the underlying fear that one day it could just stop shining. This is why total eclipses were deeply terrifying in so many cultures; they represented a temporary loss of sunlight during daytime, a temporary defeat of the Sun god to the demons of darkness. And since this "defeat" does happen for a short span of time, why not permanently? These fears are deeply ingrained in our collective mind, even if now they are redressed in poetic rather than in religious awe, as any witness to a modern-day eclipse can testify, and as I hope to have convinced you with my own testimony. The feeling of exhilaration as we watch the sunlight reemerge from behind the darkened lunar disk is overwhelming and irrational. To this, I will now add the also beautiful, but rational, description of how stars shine, and how one day the Sun will indeed cease to be able to support life on Earth.

Solar Chemistry

Many Greek philosophers brushed aside Sun worshiping as superstitious nonsense; Aristotle championed the view that the Sun, like all other celestial luminaries, was made of a fifth kind of matter, different from the four basic "elements" that composed all things material here on Earth—air, water, earth, and fire. This was the "quintessence," or ether, the matter that was no matter, because it was not subject to the transformations undergone by usual matter. Ether was unchanging and eternal, and all things made of it naturally moved in circles; hence the circular motion of the Sun, stars, and planets around Earth. Dismantling the view that celestial luminaries and objects of Earth were made of different stuff would take over twenty centuries.

The first blow was delivered by Galileo, who in 1613 published his *Letters on Sunspots*, where he correctly argued, against a Jesuit priest named Christoph Scheiner, that sunspots were blemishes on the Sun itself and not planets orbiting close to it. Thus, the Sun was not a perfect globe of ether but had imperfections that actually changed over time; the sunspots moved about its surface like acne. In spite of Galileo's observations, not much progress was made in understanding the Sun's material composition until early in the nineteenth century, when the German lens maker Joseph Fraunhofer accidentally discovered hundreds of black lines in the solar spectrum. Up until then, it was believed that the

Sun had a perfectly continuous spectrum, obtained by making sunlight pass through a prism, the colors smoothly shifting from the deepest violet to red, as with a rainbow. The black lines represented colors that were mysteriously missing, as if some capricious god were filtering colors selectively, perhaps to embroider a new robe for Amaterasu; at the time, this was as plausible an explanation as any. By the mid-nineteenth century, it was clear that gaseous samples of chemical elements, when heated up, glowed very selectively as well, each element shining with a specific set of colors, its own spectrum. The heat that warms the gas is reprocessed and reemitted in personalized rainbows— hydrogen with large amounts of violet; sodium, of yellow; neon, of reds and oranges; and so forth. The reverse was also shown to be true. When light with a continuous spectrum passes through a cool gas sample of a given chemical element, the element absorbs the exact same colors it emits when it is hot: sodium absorbs yellow, neon reds and oranges, and so on. When the element is "cold," it thus eats certain colors, and when it is hot, it emits those same colors, as figure 28 in the insert illustrates.

On the basis of these observations, it became possible to assemble a catalog of spectra belonging to different chemical elements. This catalog could then be used to explain the missing lines in the solar spectrum. If we suppose, quite reasonably (and correctly), that the solar interior is much hotter than its outer layers, the continuous spectrum generated within would have certain lines "eaten away" by whatever chemical elements are found in its cooler outer layers. In reality, the situation is a bit subtler; the Sun can be thought of as having several different layers, like a giant onion (see figure 8). The light that we see (i.e., the radiation emitted by the Sun in the visible portion of the spectrum), which defines the solar "surface," comes from the photosphere, where the temperature is about 6000 degrees Kelvin.* Even though the photosphere is made up mostly of just a few chemical elements, each with its very own signature spectral lines, the total spectrum at the photosphere is continuous, with no dark lines. At 6000 degrees, all chemical elements are in their gaseous state. However, the atoms of these elements are very tightly packed, since the Sun's average density is 1.4 larger than that of liquid water. So, we have a gas, but a very congested one. On top of the congestion, the high temperature causes the atoms to jiggle wildly; at

---

* A quick note on temperature scales. In physics, we don't like the unnatural Fahrenheit scale. Instead, we use mostly the Kelvin or the Celsius scale. The Celsius scale sets the freezing point of water at 0 degrees and the boiling point at 100 degrees, a very reasonable choice. To go from the Celsius to the Kelvin scale, you simply add 273 degrees. Thus, 0 degrees Celsius means 273 Kelvin, while –273 Celsius means 0 Kelvin, the so-called absolute zero.

the atomic and molecular levels, temperature is understood as motion, and the higher the temperature, the faster the atoms move and collide with each other. These fierce collisions destroy the coherence of the individual spectra, creating a jumbled optical mess, which shows up as a continuous spectrum with no dark lines.

As the light from the photosphere passes through the thinner and cooler chromosphere, selective filtering finally occurs, producing the dark lines observed by Fraunhofer. Comparing the different spectral lines in the "spectral catalog" with the missing lines in Fraunhofer's solar spectrum, we can infer which chemical elements make up the Sun's cooler outer layers. In the early 1920s, the Indian astrophysicist Meghnad Saha played a key role in solving the extremely complicated puzzle of the spectral lines, obtaining the chemical composition of the Sun. But Saha went a step further and helped determine not only the chemical elements that make up the Sun but their proportions as well. In order to understand how he and others did this, it is important to recall how atoms are organized.[4]

The first working model of the atom was proposed by the Danish physicist Niels Bohr in 1913. Although it relies on several simplifying assumptions, it does incorporate some of the essential features of modern atomic theory, and it

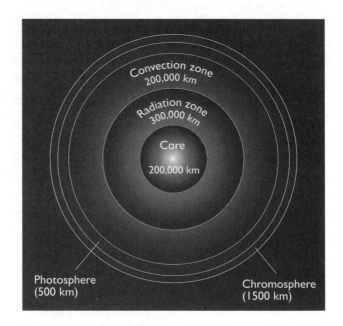

FIGURE 8: *The many layers of the Sun and their respective radii. (Not shown are the outlying transition zone and the solar corona. Figure not to scale.)*

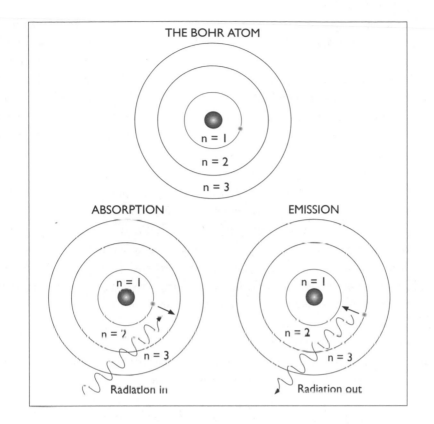

FIGURE 9: *Emission and absorption of radiation (photons) according to Bohr's model of the atom. To jump a level or more, the electron must absorb a photon of the precise energy difference between the two levels. To go down a level or more, the electron releases a photon of the energy difference between the two levels.*

will serve us well. There is a heavy and very small nucleus, formed by the positively charged protons and the electrically neutral neutrons. (Bohr didn't know about the neutrons in 1913, since they were discovered only in 1932, by Sir James Chadwick.) Negatively charged electrons move around the nucleus at very specific orbits, or levels. They cannot be in between two orbits, just as we cannot be in between two steps of a ladder (see figure 9). The lowest orbit, the first rung of the ladder, is the "ground state" of the atom, while higher rungs are called excited states. Since the positively charged nucleus attracts the negatively charged electrons, you must supply energy to the electron in order to move it to higher orbital levels. This energy comes in little packets of electromagnetic radiation called photons; each color of light is made of photons with a characteristic energy. The same is true for invisible types of electromagnetic radiation, such as x rays, infrared, ultraviolet, radio, and so on. In order for an electron to move up a rung, it needs energy. It gets this energy by absorbing a photon of the exact energy difference between the two rungs; the electron "eats" the photon, becomes more energetic, and happily jumps up an orbit. Sometimes the required photon is in the visible part of the spectrum, and sometimes it is not. The converse is also true; in order for an excited electron to come down a rung, it spits out a photon with energy equal to the energy difference between the two rungs. It is also possible to have jumps or, better, transitions between faraway orbits, although some restrictions apply. As a general rule, transitions over many rungs require more energetic photons not necessarily within the visible part of the spectrum, such as ultraviolet or x-ray photons.

We now understand the origin of the emission and absorption spectra of different elements in a much deeper way; if you shine white light (continuous spectrum) at a given element, there will be photons of all possible energies from red to violet (and the invisible ones) hitting the atoms of the element. Since each element has a fixed number of protons and electrons, it will also have a unique set of possible orbits for the electrons; each element has its own ladder, with rungs of different heights. So, the electrons of each element will "eat" those photons corresponding to the very specific transitions between its orbits, letting the other photons pass through. The electrons then climb up in their orbits, and photons end up being missed at the other end—the dark lines typical of absorption spectra. If you instead have an excited atom and let it be, it will eventually relax down to its ground state by emitting photons corresponding to the specific jumps of its electrons down their very own ladder of orbits—the emission spectrum.

Now, if a very energetic incoming photon hits an atom, or if atoms are colliding at high enough temperatures (as in the photosphere), it is possible for the atomic electrons to be not only pumped up to higher energy levels but kicked out altogether; in this case, we say the atom gets "ionized." For each electron that gets kicked out, the atom—now called an ion—gains a net positive charge. In the case of hydrogen, with its single electron orbiting a single proton, the ionization leaves just a proton zooming around. Ionized hydrogen does not have an emission spectrum, because there is no electron to eat up passing photons. Now we can go back to Saha and the chemical composition of the Sun. Saha realized that at 6000 degrees Kelvin, practically all hydrogen atoms are ionized and should not produce any hydrogen-specific absorption lines in the Fraunhofer spectrum. And yet, there they are! Saha correctly deduced that the only explanation for this is that the Sun has huge amounts of hydrogen so that, even though most are ionized, the ones left over eat up enough photons to make up the observed dark lines typical of hydrogen absorption. He then repeated this argument for other elements, with different numbers of electrons and requiring different ionization energies. For example, to strip one electron from the helium atom (two protons and two electrons) takes a little less than twice the energy to ionize hydrogen, while to strip the other takes a bit over four times as much. To make things even more exciting, ionization changes the spectra of elements; thus, the spectrum of He+ (helium less one electron) is different from that of He. (Or of He++, helium less two electrons. In fact, should He++ have an absorption spectrum?)*

The studies of Saha and others revealed that the Sun is made up of 91.2 percent hydrogen and 8.7 percent helium (in today's numbers). The leftover 0.1 percent is composed of several other elements, including oxygen, carbon, nitrogen, and silicon. The truly remarkable point here is that the chemistry of the Sun is as familiar to us as the chemistry of Earth; the same chemical elements we find here are found there, albeit in very different proportions. As we have seen, large amounts of hydrogen and helium are found in Jupiter and the other giant gas planets, but not in the inner rocky planets, like Earth.

Establishing the chemical dominance of hydrogen in the Sun offers a very important clue as to how it shines. Although the dark lines in the solar spectrum give direct evidence only of the composition of the Sun's absorbing outer layers, it is widely thought that the above numbers hold true for most of the Sun, with the exception of its inner core. And that's where the real action is.

Solar Alchemy

We start by going back to our description of the Sun as having several layers, like a giant onion (figure 8). The three innermost layers are by far the largest ones, starting with the solar core, which has a radius of 200,000 kilometers. Here the solar engine works at full blast, a true nuclear hell, where temperatures reach 15 million degrees Kelvin. The enormous energy generated, packed in highly energetic photons (mostly gamma rays), flows outward through the next two layers, the radiation zone and the convection zone. The different names stem from the different mechanisms by which energy is transported outward through these two regions. In the inner part of the radiation

---

* Of course not! There are no electrons to absorb any photons, right?

zone, as the name indicates, the heat from the core travels outward through radiative processes: the temperatures near and around the core are so high that all atoms are ionized through their fierce collisions with one another and do not have any electrons to absorb the outgoing photons and jump to higher orbits. Instead, the ionized electrons move freely about, colliding with the outgoing photons and slowing them down, as in an obstacle course. As we move outward from the center, the temperature drops and more and more atoms manage to retain their electrons and thus can absorb photons. In fact, by the end of the radiation zone, some 300,000 kilometers away from the core (see figure 8), *all* photons get captured; its outer edge is almost perfectly opaque [photons get blocked]).

Since we do see radiation coming out of the Sun, this cannot be the end of the story. Where the radiation zone ends, the convection zone begins; here heat is transported outward very much as in a boiling pot of soup; hot gas "floats" upward, cools off, and sinks back down, in a cycle of heat exchange, or convective motion, from the solar interior to its surface and back down. The sizes of these rolling convection cells vary from 30,000 kilometers deep in the convection zone to a "mere" 1,000 kilometers at its end. Once they reach this size, some 200,000 kilometers away from the radiation zone, the gas is so rarefied that it cannot support convective motion any longer; photons begin to radiate outward, transporting the heat away from the Sun; this is the photosphere that we see. It may take a photon millions of years to travel from the Sun's inner core to the photosphere. From there, it takes only 8.3 minutes for it to catch our eyes here on Earth.

Now that we know how heat escapes the solar interior, we focus on how it is actually produced. The process by which the Sun generates its amazing amount of heat is known as nuclear fusion; the Sun's luminosity or power, the total energy it radiates *per second*, is 4×10^{26} watts, that is, 4 followed by 26 zeros watts, equivalent to 100 billion 1-megaton nuclear bombs. Compare this with your living room light bulb, at 100 watts. More than that, the Sun has been generating this amount of energy for about five billion years and will continue to do so, at a fairly steady rate, for another five billion years. Where does all this energy come from?

To answer this question, we leave electrons behind and plunge into the atomic nucleus. The state of matter that exists in the solar core is known as a plasma, made of atoms stripped of all their electrons as a result of their violent mutual collisions—that is, bare nuclei and electrons flying around independently. In its solid state, matter has its atoms arranged in regular crystalline structures, such as cubic or pyramidal shapes. If we try to squeeze a solid into a smaller volume, the electric repulsion between electrons from its atoms will make them repel each other fiercely, giving a solid its rigid structure. In plasmas, since the electrons are all gone, the nuclei can be squeezed into distances ten thousand times smaller, and still move with relative ease. This intimacy entices nuclei to fuse, creating different kinds of nuclei. These processes, where two or more atomic nuclei fuse into another nucleus, are known as *fusion* nuclear reactions. To understand the basic rules of fusion, we must first learn about atomic accounting.

Each chemical element is characterized by the number of protons in its nucleus, known as the atomic number, and represented by the letter Z. Thus, hydrogen, the simplest, has only 1 proton (Z=1), helium has 2 (Z=2), lithium has 3 (Z=3), uranium has 92 (Z=92), and so forth. The other component of the nucleus is the neutron, which can come in different numbers. For example, whereas normal hydrogen has no neutrons in its nucleus, it can appear in variant forms with 1 or 2 neutrons. These variant forms are called isotopes. To make things simple, chemists like to represent elements and their many isotopes by specifying the number of protons plus neutrons (the atomic weight, A) in their nucleus. The notation is as follows: ^1H is normal hydrogen, with 1 proton (A=1); ^2H is its isotope known as deuterium, with 1 proton and 1 neutron (A=2); ^3H is its isotope tritium, with 1 proton and 2 neutrons (A=3). One more example: ^{235}U is a rare isotope of uranium with 143 neutrons. (You get this by subtracting 92, the atomic number of uranium, that is, the number of protons in its nucleus, from 235.) Although there are only 92 naturally occurring chemical elements, there are thousands of isotopes.

During the first two decades of the twentieth century, the New Zealand physicist Ernest Rutherford realized the greatest dream of the alchemists, to achieve element transmutation. Granted, he was not trying to change lead into gold, restricting himself to less profitable, but no less fundamental, transformations between chemical elements. This can be done in two ways; one is by fusing their nuclei, and the other is by breaking them apart. In either case, the reacting nuclei must come very close together through collisions, which means they must overcome their mutual electric repulsion or barrier; in other words, nuclear reactions require a lot of input energy so that the nuclei can "touch" in spite of their positive electric charges. I put "touch" in quotation marks to stress that nuclei are to be thought of as small deformable distributions of matter, more like sponges than hard billiard balls. In 1919, Rutherford announced that he had succeeded in changing nitrogen (Z=7, A=14) into an isotope of oxygen (Z=8, A=17) by bombarding the nitrogen with nuclei of helium (Z=2, A=4). This nuclear reaction can be represented as follows:

$$^{14}N + {}^4He \rightarrow {}^{17}O + {}^1H$$

As in any nuclear reaction involving element transmutation, there is a shuffling between protons and neutrons; notice how the sum of the atomic weights is preserved. (The total atomic weight A=18 on both sides.) Of the two protons from the helium nuclei, one is "eaten" by the nitrogen, changing it into oxygen, and the other is expelled (the hydrogen nucleus). This particular nuclear reaction consumes more energy than it liberates. However, with it Rutherford proved that the nature of chemical elements is not written in stone but is amenable to change, as the alchemists dreamt, so long as they are exposed to the right conditions, of which the alchemists couldn't have dreamt.

Soon after Rutherford's initial nuclear transmutations, it became quite clear that nuclear reactions could liberate vast amounts of energy. This is true for nuclear breakdown reactions (or fission reactions), such as those occurring in atomic bombs and nuclear reactors, and for nuclear fusion reactions, as in hydrogen bombs and the interior of stars. Even though we are interested in fusion reactions here, I cannot let you move on without at least briefly explaining fission reactions. After all, they gave us both nuclear power stations and the infamous power to destroy life on Earth many times over, irreversibly changing the collective history of humankind. In a typical fission reaction, a very heavy nucleus is hit by a neutron and split into two medium-size nuclei, liberating a lot of energy in the process. Two isotopes are particularly important, uranium 235, which we met above, and plutonium 239. The reason for their importance is that the splitting of these heavy nuclei produces not only medium-size nuclei

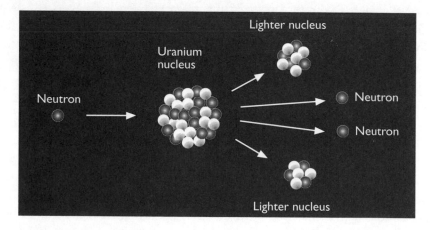

FIGURE 10: *Schematic diagram of nuclear fission. A neutron hits a uranium nucleus, splitting it into smaller nuclei and creating a few extra neutrons.*

but also a few free neutrons. These neutrons act like bullets and hit other heavy nuclei, splitting them and creating more neutrons that will split more nuclei, and so on. This is what we call a chain reaction; a very fast one becomes an atomic bomb, whereas slower ones generate energy in a controlled way, as in nuclear power stations.[5]

The secret behind the power liberated in fusion reactions is the transformation of matter into energy. As Einstein proposed early in the 1900s, there is a relationship between the mass (m) of an object and its energy (E), encapsulated in the famous equation $E = mc^2$, where c is the speed of light in empty space, 300,000 kilometers per second. What does this formula, perhaps the most popular but also the most misquoted in all of physics, actually mean? It means that we can think of mass as stored energy; in the same way that we can store energy in a spring by squeezing it and then releasing the stored energy by letting the spring go, any massive body can have its mass released as energy—in the form of electromagnetic radiation—under the proper circumstances. Fortunately, those circumstances are exotic enough that we don't see people or dogs turning into a flash of highly energetic photons; if we did, it would be the last thing we would ever see; if we could convert 1 kilogram of matter into energy in one second, it would generate about 10^{17} watts of power! However, the conditions under which mass turns into energy are routinely created in experiments involving highly energetic particle collisions and, of course, in the interior of stars.

The converse is also true, "energy" can turn to matter; it is possible for highly energetic photons, the bullets of electromagnetic radiation, to spontaneously create particles of matter. Because photons are electrically neutral, and electric charge is conserved in particle reactions, the material products of their demise must also add up to total zero charge. A typical example is the transformation of photons into electron-positron pairs. Positrons are electrons with positive charge, also known as the electron's antiparticle. According to the laws of particle physics, every particle of matter has a partner antiparticle of "antimatter"; the electron has the positron, the proton the antiproton, and so on. Thus, a typical matter–energy transformation can be represented as

$$\text{particle} + \text{antiparticle} \leftrightarrow \text{photons (energy)}$$

So, if you shook hands with your antibeing, you would both disintegrate into a burst of gamma-ray photons, which would probably destroy a big chunk of the United States. For electron-positron encounters, the amount of energy is more modest since the masses involved are much smaller. In any case, I want to reassure you that there is nothing to fear—you won't be hit by a wandering bunch of antiparticles. Matter is much more abundant than antimatter in the universe; antimatter particles are extremely rare, appearing profusely only at the heart of very energetic particle collisions, such as those promoted by modern particle accelerators. Had it been otherwise, we would not be here discussing solar physics. In fact, we wouldn't be here at all.

The basic law controlling these mass–energy transformations is that the total amount of mass and energy must be the same before and after; if some mass is lost, it turns up as energy and vice versa, but the total sum of the two (in appropriate units) must be conserved. Fusion reactions make full use of this general mass-energy conservation; the mass of the fused nucleus is smaller than the masses of the fusing nuclei, with the difference being converted into some sort of energy. Thus, a general fusion reaction can be written as follows:

$$\text{nucleus-1} + \text{nucleus-2} \rightarrow \text{nucleus-3} + \text{energy}$$

This surplus of energy powers the Sun and other stars, although the types of fusion reactions will depend on the stars' core temperatures, which are determined by their masses. For modest stars like the Sun, the core temperature can reach about 15 million degrees Kelvin, high enough to initiate the so-called proton-proton chain, the first rung in the nuclear fusion buildup.

The simplest of all nuclei is the hydrogen nucleus, with its single and lonely proton. Clearly, if two protons are to fuse into something else, they must overcome their electric repulsion. But even if they are moving extremely fast, so as to collide hard with one another, what makes them stick? At these very small distances, another fundamental force of nature comes into play, the *strong nuclear force*. It is this force, which acts only within nuclear distances, but is about a hundred times stronger than the electric repulsion between protons, that fuses nuclei together. Incidentally, it is also the strong force that glues the atomic nucleus into a wholesome lump, with all its repelling protons and electrically indifferent neutrons. The strong force's short range explains why there are only ninety-two stable atoms in nature; any more protons, and the strong force could not overcome their electric repulsion and glue them together within a small nuclear-size volume.

To recap, nuclear fusion takes advantage of the matter-energy conservation law, as matter is converted into energy during the fusion process. In order for nuclei to fuse, they must be moving at extremely high velocities so as to overcome their mutual electrical repulsion, and get sufficiently close to allow the strong nuclear force to perform its transmutation trick. The first rung in the fusion ladder is also the simplest, two protons fusing into a nucleus of the

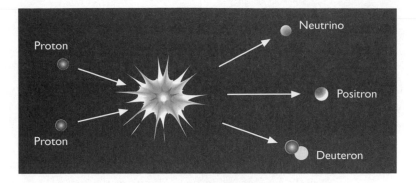

FIGURE 11: *Two protons fuse into a deuteron, the nucleus of the hydrogen isotope deuterium. A neutrino and a positron are also created during the fusion process.*

hydrogen isotope deuterium (^{2}H), as shown in figure 11. The "fusion" here actually involves the transmutation of a proton into a neutron.

There are two other reaction products: a positron, the "antielectron" we encountered above, and a particle called neutrino, which, like the photon, has no electric charge (hence the name, "little neutron") and has at most a tiny mass (we still don't know for sure), at least millions of times smaller than the electron's, if any. Neutrinos have the remarkable property of hardly ever interacting with other particles; in fact, they easily travel across the Sun and out to space at the speed of light or very close to it. Not so for the positrons. Recall that the plasma at the solar core is rich both in hydrogen nuclei (the protons) and in electrons. Therefore, the freshly minted positrons meet a very fast death as they collide with their matter partners, disintegrating into highly energetic gamma-ray photons. In short, the first rung in the nuclear fusion chain consists of two protons fusing into a deuteron (the nucleus of deuterium) and producing gamma-ray photons and neutrinos. These two particles carry away the energy surplus of the fusion process.

The next rung in the ladder is to fuse the newly made deuterons with another proton to make the next-heaviest isotope, ^{3}He, the lightest isotope of helium, with two protons and one neutron in its nucleus:

$$^2\text{H} + {}^1\text{H} \rightarrow {}^3\text{He} + \text{energy}$$

Again, the energy comes out in the form of gamma-ray photons. The next and final step in the proton-proton chain is the actual fusion of a helium nucleus, ^{4}He. This can be accomplished in a number of different ways, but the most efficient is to fuse two ^{3}He isotopes, produced in the preceding step:

$$^3He + {^3He} \rightarrow {^4He} + {^1H} + {^1H} + energy$$

The net balance of the three-step reaction is as follows: for each ^3He we used three hydrogen nuclei (^1H), and we needed two ^3He to produce the final ^4He. This last step also produces a surplus of two hydrogen nuclei. So, we used a total of six hydrogen nuclei but got two back from the helium fusion. The total helium fusion reaction can be summarized, including the two neutrinos from step one (performed twice), as follows:

$$4(^1H) \rightarrow {^4He} + energy + 2\ neutrinos$$

The energy liberated at the core as gamma rays flows into the radiation zone, weakening along its way until it reaches the convection zone, where it is boiled up to the solar surface, radiating in the visible and infrared.

The Sun will remain in this hydrogen-burning phase, also known as the main sequence, for most of its life cycle, roughly for another five billion years. At that point, the continuous fusion of hydrogen nuclei at the core will deplete its reserves, and the Sun will have to fuse something else to generate enough energy to contain the inexorable gravitational crunching of its outer layers. This is the beginning of the helium-burning cycle, the beginning of the end for the Sun and stars of similar mass and composition. I will now tell this tale of self-cannibalism, where stars literally devour their own entrails in order to survive.

Red Giants

A star's fate is sealed by its mass; very massive stars, much more massive than the Sun, must struggle much harder to contain the inward pull of their own gravity. Like any star, their only weapon is their own matter, which must burn by fusion at a furious pace to release the enormous energy needed to detain their collapse. In the end, their large masses compromise their longevity; the most massive stars live only a few tens of millions of years before encountering a violent death. We met them before, during our discussion of the formation of the solar system, in chapter 3; their remains sprinkled our neighborhood with heavier elements and the small grains needed for the formation of planetesimals. They may also have caused the initial gravitational instability of the progenitor hydrogen cloud, which contracted to become our solar system. At the other extreme, we find small cold stars called "red dwarfs," which burn so slowly that they may live for trillions of years. The Sun is in between these two extremes, a star that can fuse hydrogen into helium through the proton-proton chain for about ten billion years. The question, then, is what happens after these ten billion years of fairly steady burning.

In order to follow the evolution of a star through its final stages, it is useful to keep in mind the interplay between the two main forces determining its fate. While the attractive force of gravity wants to squeeze the star into the smallest possible volume, the heat generated by this very squeezing creates a counter-pressure that balances the star's collapse. So, in order to survive as a stable hot ball of gas, the star must keep producing heat, which comes from fusing hydrogen nuclei at its core. Clearly, this dangerous game of consuming its own entrails for survival cannot last forever. For now, we will focus on what happens with stars like the Sun, which have fairly low masses. As we will see in the next chapter, the story changes quite dramatically for stars of masses larger than roughly eight times that of the Sun—the realm of supernova explosions and their exotic remnants, neutron stars and black holes. But before we get to that, we will use modern astrophysical ideas to foretell the Sun's tragic fate, which, as the Egyptians, Incas, Japanese, and many other peoples have known for many centuries, is of great consequence to our own.

For about 10 billion years, hydrogen fuses into helium, creating enough energy to balance gravity's crush. As helium becomes more abundant at the center, hydrogen becomes rarer. At its present age, the Sun has converted only about 5 percent of its total mass into helium. As this amount increases, so will the Sun's luminosity, that is, its total power output. In 1.1 billion years, the luminosity will be 10 percent larger than today; in 3.4 billion years, 40 percent larger. This extra amount of power will have serious consequences for Earth's climate, creating first a "moist greenhouse"effect and then a true "runaway greenhouse" effect. Possibly, clouds may delay these effects somewhat, but the prospect for life on Earth beyond roughly 1 billion years is quite grim. The rapid increase in temperature will melt the polar ice caps, causing the oceans to rise and flood all coastal areas. The temperature will rise some more, and the oceans will boil, thickening the atmosphere with dense clouds of steam. Lurking behind the thick vapor, a faint Sun will seem to be mocking an Earth that looks much like Venus.

The amount of helium will keep growing, and in five billion years it will reach a crisis level, as practically no hydrogen nuclei (protons) will survive at the Sun's core; most hydrogen burning will happen in a shell surrounding the core. This is when serious trouble begins; if helium fused into something else, the Sun's doom would be postponed, since the heat released would act to stop any further contraction. The problem is that helium nuclei, with two protons, are harder to fuse because of their larger mutual electric repulsion. In fact, the fusion of helium requires temperatures on the order of hundreds of millions of degrees, as opposed to the ten or so million of hydrogen fusion. Thus, when the core becomes helium dominated, it is too cold (by a factor of ten) to induce helium fusion; gravity, of course, takes advantage of this lowering of the guard, squeezing the helium core further. This causes the temperature at and around the core to increase, heating up even more the hydrogen in its surrounding shell, which begins to burn at a furious pace. The star enters a schizophrenic phase, where the "cold" helium core continues to shrink and heat up while its hydrogen envelope burns and expands, pushing the nonburning layers outward (see figure 12). This is the red giant phase, when the Sun will grow to one hundred times its normal size (the "giant" in red giant). This growth will cause its surface temperature to decrease, as a result of the thinning of the gas in its outer layers, which will radiate in the cooler red end of the visible spectrum (the "red" in red

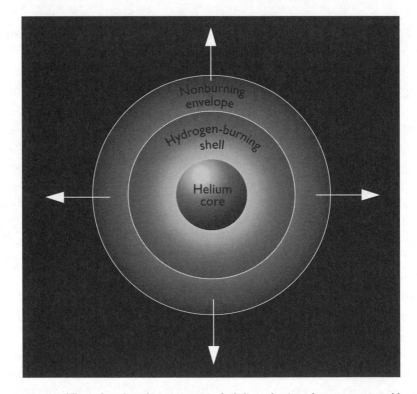

FIGURE 12: *The path to the red giant stage. As the helium-dominated core contracts and heats up, hydrogen burns even more furiously in the surrounding shell. The enormous pressure generated there pushes the star's nonburning envelope outward.*

giant). A powerful red giant can be seen with the naked eye in the constellation Orion, the one easily identified by the "belt" of three aligned stars. The star corresponding to Orion's "right shoulder" is a red supergiant called Betelgeuse, one of the brightest stars in the night sky. (We will soon see what the "super" stands for.) After a careful look, you will notice that it does indeed twinkle red.

When the Sun reaches the peak of its red giant phase, about one hundred million years after depleting all hydrogen at the core, its size will engulf Mercury's orbit, and it will blast about 30 percent of its mass into space. To make things worse for our distant descendants, if any are still around then, the Sun's luminosity, the power it generates, will become thousands of times larger than what it is today. If there was any form of life that survived all the pre-red-giant-phase instabilities and greenhouse effects, none will remain after it. Eventually, the atmosphere will also boil away and stars will shine at daytime side by side with the Sun; what we call the sky will be literally gone, and the planet (let's not call it Earth anymore) will be bombarded with lethal ultraviolet radiation. Rocks will melt and the planet's surface will boil like soup, as the Sun's outer layers brush close by. We can only hope our descendants will be smart enough to have left Earth well before its fiery end, when hell descends from heaven, an apocalyptic image worthy of Revelation.

But the swollen red Sun is not about to give up. It will fight the inward pull of gravity until the end, with the only weapon it has   the fusing of whatever is at its core.

While the outer layers of the star expand outward driven by the enormous heat generated by the hydrogen-burning shell, its helium core, devoid of any pressure-producing heat, continues to shrink slowly. When the core density reaches a staggering 100 million kilograms per cubic meter (that is, a cube one meter a side weighs 100 thousand tons) the temperature finally rises to the 100 million degrees Kelvin required for helium to fuse into carbon and reignite the inner solar fire. Carbon fusion happens in two steps; first two helium nuclei (^{4}He) fuse into the highly unstable beryllium 8 (^{8}Be). Before this isotope of beryllium can break down, it encounters another helium nuclei and fuses with it into carbon (^{12}C). The two-step fusion reaction can be schematically represented as follows:

$$^4He + {}^4He \rightarrow {}^8Be + energy$$
$$^8Be + {}^4He \rightarrow {}^{12}C + energy$$

Because helium 4 nuclei were traditionally called alpha particles, this fusion reaction, using three helium 4 nuclei, is known as the triple-alpha process.

But carbon fusion is not the only process triggered by core densities of hundreds of millions of kilograms per cubic meter; recall that during the hydrogen-burning stage the core was rich not only in hydrogen nuclei but also in electrons stripped from their parent atoms. These electrons finally get fed up with all this squeezing and start to react. According to quantum physics, every subatomic particle is characterized by a set of numbers that label all its relevant properties. Called quantum numbers, they include information about the particle's energy (for example, in which energy level of the atom the electron is), its angular momentum (how fast it rotates about the atomic nucleus), and its intrinsic angular momentum or spin, which, using a rough image, tells how the particle rotates about its own axis, compared with a fixed direction in space, say, the up direction. (Picture a top that can spin only at fixed angles with respect to the up direction; each angle specifies a rotation state of the "quantum top," or particle.) Once this set of numbers is given, we can describe what the electron is doing—that is, we can specify its quantum state.

One of the problems with Bohr's formulation of the atom—where, as we have seen, electrons move about the nucleus in discrete orbits—is that it didn't explain what prevented all the electrons from just bunching together at the lowest energy state, closest to the nucleus. If electrons could aggregate there, all higher-energy atomic levels would be empty and atoms would not be able to combine with other atoms to form molecules; among other things, the chemistry of life would not have been possible. It turns out that electrons are very antigregarious particles and cannot coexist at the same quantum state. For atoms, this means that electrons fill up the available discrete orbits, like marbles placed in pairs on the steps of a staircase; if they are squeezed too close to each other by some outside force, they respond by counterimposing a pressure known as electron degeneracy pressure. It is this pressure that will help support the helium core against further collapse. This antigregarious behavior of the electrons and many other particles, including protons and neutrons, is known as the Pauli exclusion principle, after Wolfgang Pauli, who proposed it in the mid-1920s.

As with a spring mattress, the heavier the sleeper, the harder the springs must work to resist the downward pressure caused by his weight. Likewise, the degenerate electrons will try to resist the gravitational squeezing of the helium core by the outer layers of the star. To do this, they will have to move faster and faster—the higher their speed, the higher the counterpressure they can produce. Think of water flowing out of a hose and how the faster it flows, the higher the pressure it exerts on objects. There is a limit to how much gravitational squeezing degenerate electrons can stand; at some point, their speeds will approach the speed of light and, as we will soon see, other effects come into play. For now, though, the important property of this electron pressure is that it does not depend on temperature; as helium fusion begins at the agonizing Sun's core and temperatures start rising, most of the gravity-opposing pressure comes from degenerate electrons, as opposed to thermal pressure. This changes the physics at the core quite dramatically: during normal burning, when the star's core heats up because of fusion, it expands, cools, and contracts until a reasonable fusion rate is found that properly balances the gravitational squeezing without burning too much fuel. A thermostat in an air conditioner is a reasonable analogy, because it turns the engine on and off to keep the temperature at a set value. However, since degenerate electrons do not respond to temperature changes, when helium fusion raises the core temperature, there is no corresponding expansion and cooling, as if the star's thermostat were broken; the temperature shoots up, setting a bomb-like pace of helium burning, known as a helium flash. After a few hours of out-of-control helium burning, the temperature rises to a point where thermal pressure finally begins to dominate again. The star's thermostat clicks back on, and helium burns to carbon at a more moderate pace, albeit still very fast, for a few tens of millions of years.

Again, this relative peace cannot last for long. The now carbon-rich core gets depleted of helium, and the star falls into a similar cycle: the carbon core is too cold for carbon fusion and starts to shrink under its own gravity, while its surrounding helium shell keeps burning, being itself surrounded by the hydrogen-burning shell that previously enclosed the helium core (see figure 13).

Just as before, the shrinking of the carbon core causes an increase in temperature, which heats up even more the helium- and hydrogen-burning shells. All this heat causes the outer layers of the star to expand outward, in a dramatic repeat of the red giant phase. But the temperatures are now much higher than before, and the convulsing star grows even larger than before, swallowing everything past the orbit of Mars. Earth will be no more.

If the core temperature climbed to about 600 million degrees Kelvin, carbon could start fusing into heavier elements, and the release of heat would support the star's collapse for a bit longer. But as the core density reaches a staggering 10 billion kilograms per cubic meter—a cherry-size chunk of this stuff would weigh one ton on Earth—the electrons once more start to react against any further squeezing, their pressure stopping the contraction of the carbon core. As a result, the temperature never rises enough for carbon fusion to begin in earnest: the gravitational contraction of the carbon core gets balanced by the electron degeneracy pressure, a quantum effect balancing an astronomical object, a true

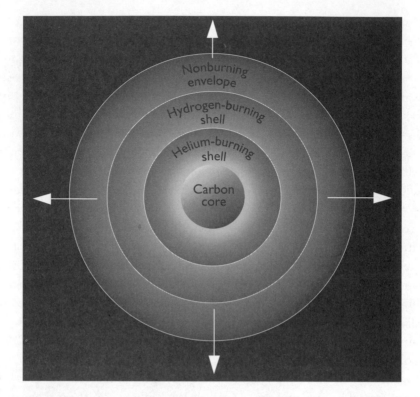

FIGURE 13: *The path to a red supergiant. The helium at the core fuses into carbon. The shrinking carbon core is surrounded by three layers: the helium-burning layer, the hydrogen-burning layer, and an expanding nonburning envelope.*

matching of the physics of the very big and that of the very small. Although a bit of oxygen does get fused at the inner edge of the helium-burning shell, the star is pretty much spent now.

The Shedding of the Veil

The story is not yet over. It is hard to imagine how an object with core temperatures of about 300 million degrees Kelvin, densities of billions of kilograms per cubic meter, and surrounded by helium and hydrogen-burning shells would just rest in peace. As the core adjusts to its electron-supported equilibrium, the burning at the inner shells becomes very unstable. A series of helium flashes occurs in the inner helium shell, liberating so much energy that the star literally starts to pulsate violently, its outer layers becoming looser with each convulsion. After about one million years of this agony, the star expels its outer layers into space at speeds of tens of kilometers per second.

The "Sun," or what's left of it, has now two separate parts; a very hot carbon core, surrounded by a thin layer where helium is fused into carbon and oxygen, and a thin, approximately spherical veil of the cooler ejected matter, which covers roughly the size of the solar system. (The more technical term is "envelope," but I prefer the image of a veil.) This veil of glowing gases, one of the most beautiful sights in the universe, is known as a planetary nebula (see figure 29 in insert).

Of course, planetary nebulae have nothing to do with planets; the term originated in the eighteenth century, when astronomers identified these colorful objects with low-resolution telescopes and were tricked into associating them with colorful planets or, as in Saturn, with rings around them, the veils of other planets. As time goes by, the ejected veil moves farther away from the core, enriching the interstellar medium with hydrogen, helium, carbon, and oxygen. A true cosmic recycler of matter, the dying star sows the seeds for new stars to be born, echoing Kant's Phoenix-like vision of death and rebirth of celestial objects.

While the veil ventures into outer space, the leftover carbon core, unable to fuse anything else, starts to slowly cool off. It is a small and compact object, roughly the size of Earth and with a mass about half that of the Sun; its stability against further gravitational contraction is guaranteed by the stubborn degenerate electrons. The lingering high temperatures make the core's surface shine white hot, although because of its small size, its luminosity (total power generated) is quite small; for these two reasons, small size and white-hot surface, these compact objects are known as white dwarfs, hot embers fading after a cosmic fire. The cooling of white dwarfs takes billions of years, a time comparable to the age of the universe; as the surface temperature drops, its glow becomes increasingly dim, until it finally slides into dark oblivion.

This gloomy scenario is reserved for stars of masses comparable to the Sun's. Larger-mass stars will crush the electrons at the core to such an extent that not even their powerful degeneracy pressure will be able to halt the star's gravitational implosion. This is the realm of extreme astrophysics, of supernovae, neutron stars, and black holes, where we now turn.

CHAPTER 6

Cosmic Maelstroms

I had a dream, which was not all a dream.

The bright sun was extinguished, and the stars

Did wander darkling in the eternal space,

Rayless, and pathless, and the icy Earth

Swung blind and blackening in the moonless air.

—BYRON, "DARKNESS" (1816)[1]

Throughout human culture, in legends, myths, and fables, there is manifest a great fascination with fantastic journeys, the exploration of unknown lands, earthly or beyond. In these otherworldly realms, flesh-and-blood heroes meet gods, dragons, and unicorns, battle evil spirits, and descend into the fiery pits of hell. Moses climbed Mount Sinai to meet God; Jason, the leader of the Argonauts, while looking for the fabled Golden Fleece battled fire-breathing bulls and sowed a dragon's teeth; Hercules, one of Jason's crew members, descended into Hades and captured Cerberus, the three-headed dog that guarded its entrance; King Arthur and the Knights of the Round Table searched for the Holy Grail; and so on. Each of these fantastic journeys has an element of self-transcendence, a contact with a higher realm of being, where the real mixes with the unreal. The hero returns deeply transformed by his experiences, not the flesh-and-blood person that started the voyage but a demigod, a leader, one who has acquired some secret knowledge not shared by common mortals, one who inspires higher morality. It is the fantastic element of these narratives that permits the deep transformation of the self that the traveler undergoes. But fantasy not only transforms; it also captivates. We, the audience, immerse ourselves in the heroes' magic and, in so doing, also become heroes, learning as we marvel; the narratives symbolize our search for knowledge and meaning, which may be revealed to us if we are brave enough to probe the unknown, to behave like heroes.

It is not very difficult to expand this argument to include the search for scientific knowledge. Scientists also wrestle with the unknown in search of enlightenment; in order to expand scientific knowledge, we must probe into unknown territory, a process of exploration that oftentimes requires great intellectual courage. Like the mythic heroes, we transform ourselves through our search, learning to face the awesome creativity of nature with deep humility.

In this chapter, we will embark on a fantastic journey dictated not by the heroic legends of the distant past but by the intellectual courage of a few astrophysicists who dared to take the path where science itself becomes a wizard, capable of creating the most amazing realities, on a par with any mythical pilgrimage. A journey into a black hole, those whirlpools of space-time, is an intellectual pilgrimage worthy of Jason or Arthur; the traveler, even if only traveling with his mind, emerges deeply transformed, his vision of the cosmos forever changed. For the universe is populated by the most bizarre creations, which demand that our very notions of space and time be deeply revised. To see why, we start our journey where we stopped last, by discussing the fate of stars heavier than the Sun, the progenitors of neutron stars and black holes.

The Unbearable Heaviness of Being

The Sun is a fairly light star. Within the main sequence—the stage during stellar evolution in which hydrogen fusion produces the energy balancing the star against its own gravity—stars can have masses ranging from one-tenth of the solar mass to twenty times it, placing the Sun in the featherweight class. We have seen how the Sun will end its life in a dramatic split into two objects, a veil of gases traveling outward at tens of kilometers per second, and a hot and dense white dwarf, where degenerate fast-moving electrons create enough pressure to contain the inexorable crush of gravity. Stars like the Sun can be thought of as cosmic recyclers of material; in go hydrogen and a bit of helium, and out go hydrogen, helium, carbon, and oxygen. During the second half of the twentieth century, it became clear that the lightest chemical elements—hydrogen, helium, lithium, and their isotopes—are mostly "primordial," that is, they were produced during the early stages of the universe's history. Since some of these elements are the basic ingredients needed to make stars, they had to come from somewhere else. But what about all the other chemical elements? Where do

FIGURE 1: *Painting of the judgment, dating from about 1025 B.C.E. Princess Entui-ny appears before Osiris, god of the dead. Maat, or Truth, with the crown made of an ostrich feather, stands at the extreme left. (Courtesy of Erich Lessing/Art Resource, New York.)*

FIGURE 2: *Luca Signorelli's depiction of the resurrection, in the San Brizio Chapel, in Orvieto. (Courtesy of Photo Scala, Florence.)*

FIGURE 3: *Luca Signorelli's powerful depiction of the Apocalypse, from the San Brizio Chapel in Orvieto. All cosmic symbols from Revelation 6:12–14 are present: the blackened Sun, the blood-colored Moon, and the raining stars* (all to the right of the picture). *(Courtesy of Photo Scala , Florence.)*

F I G U R E 4: *Depiction of Halley's comet in the Bayeux Tapestry of 1066. This comet was said to herald the arrival of William the Conqueror in 1066. Left, some people are baffled at the celestial apparition; right, an emissary is breaking the bad news to King Harold. (Giraudon/Art Resource, New York, authorized by the city of Bayeux.)*

FIGURE 5: *Sodoma's fresco in the Monte Oliveto Maggiore Abbey, Tuscany, depicting Saint Benedict exorcising a demon from a monk by flagellation.* Top, *the exorcised demon is surrounded by smoke;* center, *another demon is at work seducing an unsuspecting monk. (Courtesy of Photo Scala, Florence.)*

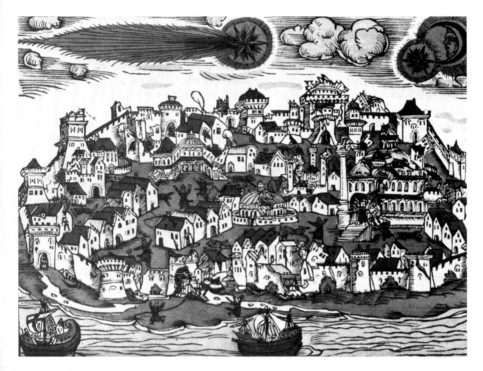

FIGURE 6: *Illustration by Herman Gall (1556) depicting Halley's comet and the havoc wreaked by its apparition.* Top right, *note the strange conjunction of the Moon and Sun, a clear sign of impending apocalyptic doom. (Reproduced from William Hess,* Himmels- und Naturerscheinungen in Einblattdrucken des XV bis XVIII Jahrhunderts *[Nieuwkoop: B. de Graaf, 1973].)*

FIGURE 7: *Luca Signorelli's* Deeds and Sins of the Antichrist, *in the San Brizio Chapel, Orvieto. (Courtesy of Photo Scala, Florence.)*

FIGURE 8: *Albrecht Dürer's* Opening of the Fifth and Sixth Seals *(1498). All cosmic symbols of doom are clearly used as elements to enhance the dramatic message found in Revelation. (Potter Palmer Collection, 1956.960.)*

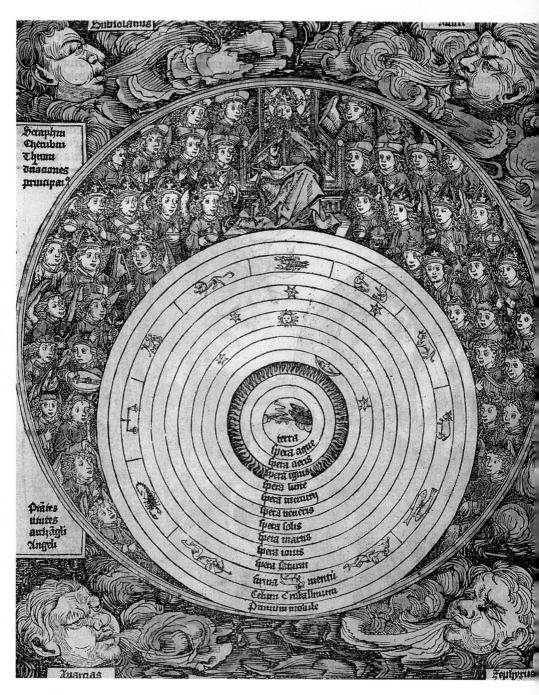

FIGURE 9: *Illustration from the* Nuremberg Chronicles *(1493), depicting Aristotle's cosmos as seen by the church. Earth in the center is surrounded by the four elements, Moon, Mercury, Venus, Sun, the rest of the planets, and the sphere of the fixed stars. The outermost sphere is the* Primum Mobile, *surrounded by the Kingdom of God. (Courtesy Adler Planetarium and Astronomy Museum, Chicago.)*

FIGURE 10: *Giotto di Bondone,* Adoration of the Magi *(1304), Scrovegni Chapel in Padua. The Star of Bethlehem* (center top) *is depicted as a comet, inspired by Halley's comet. (Copyright Alinari/Art Resource, New York.)*

Figure 11: *Diagram illustrating parallax. For a comet very far away, the angle between the two beams would be much smaller than for a comet nearby.*

FIGURE 12: *Illustration depicting a scientific debate over cometary paths, from Johannes Hevelius's* Cometographia *(1668). Left,* Aristotle *holds a diagram with comets as sublunary phenomena; Hevelius,* seated, *argues that comets come from the atmospheres of other planets and move in curved paths; right,* Kepler *insists comets move in rectilinear paths. (Courtesy Adler Planetarium and Astronomy Museum, Chicago.)*

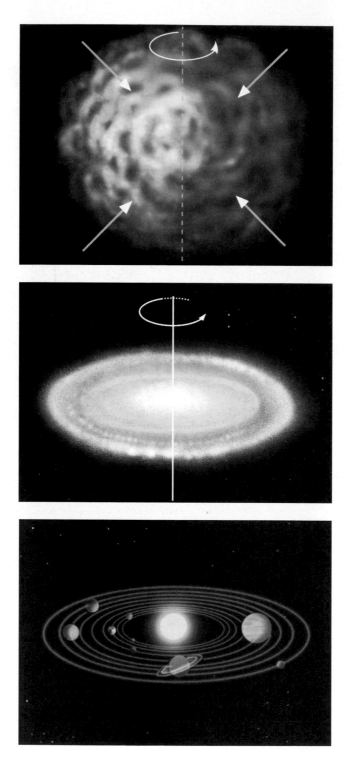

F I G U R E   1 3 :   *Sketch of Laplace's model for the formation of the solar system: (a) initial solar nebula dense in center and extending to the outskirts of solar system; (b) rings rich in gas and solid particles break off from the flattened disk because of angular momentum conservation; (c) planets form from accretion of material in rings.*

FIGURE 14: *Photo of the NGC 4414 Spiral Galaxy taken by the Hubble Space Telescope. (Courtesy of NASA.)*

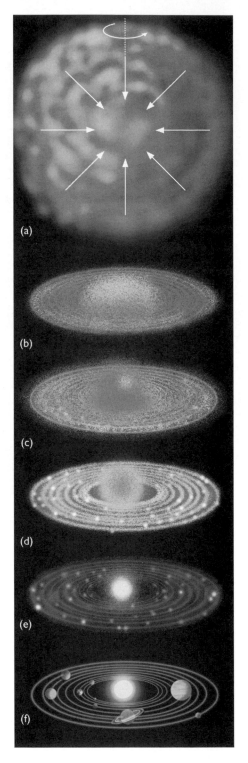

FIGURE 15: *Formation of the solar system from a contracting nebula* (a): *after flattening into a disk* (b), *the accretion of material forms planetesimals* (c–e), *which grow to become the planets* (f).

FIGURE 16: *The Meteor crater, with a diameter of 1.2 kilometers and a depth of 200 meters, carved by the impact of an iron-rich meteorite about fifty thousand years ago in the state of Arizona. (Copyright AP/Wide World Photos.)*

FIGURE 17: *A rock thrown on a pond and the outgoing waves the impact generates. Note how the disturbance caused by the rock varies with the energy of the impact.*

FIGURE 18: *An ammonite fossil, displaying its spiral structure, similar to today's chambered nautilus. (Copyright LeBlanc/Paleoplace.com.)*

FIGURE 19: *An artist's representation of the fateful impact at Chicxulub. (By Don Davis, courtesy of NASA.)*

FIGURE 20: *Photo of the flattened woods at the Tunguska site. (Copyright AP/Wide World Photos.)*

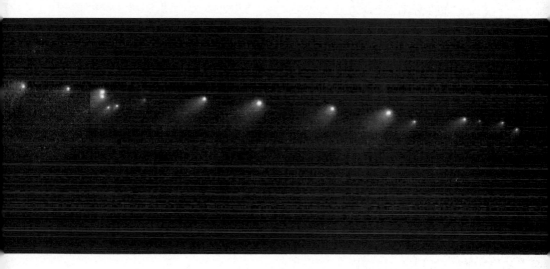

FIGURE 21: *The twenty-three fragments of comet Shoemaker-Levy 9 flying in formation toward Jupiter, as seen by the Hubble Space Telescope. (Courtesy Space Telescope Science Institute.)*

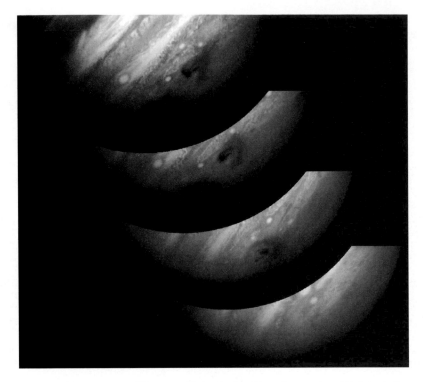

F I G U R E 2 2 : *Photomontage of fragments hitting Jupiter.* From bottom (small bright dot) up, *we see one and then two dark shock waves spreading outward on Jupiter's upper atmosphere; their sizes are comparable to Earth's diameter. (Courtesy NASA/JPL/Caltech.)*

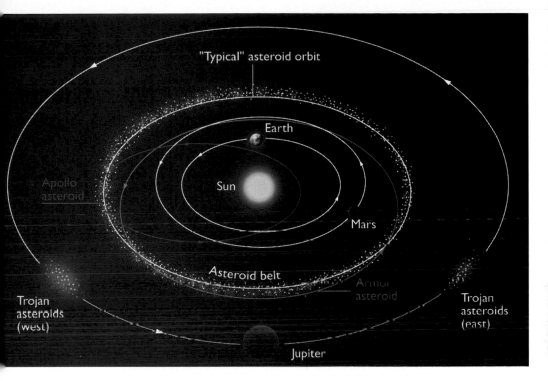

F IGURE 23: *Diagram of an Earth-crossing (or Apollo) asteroid. Also shown is an* Armor *asteroid, crossing Mars's but not Earth's orbit, and the two clusters of* Trojan asteroids, *locked into the same orbit as Jupiter's.*

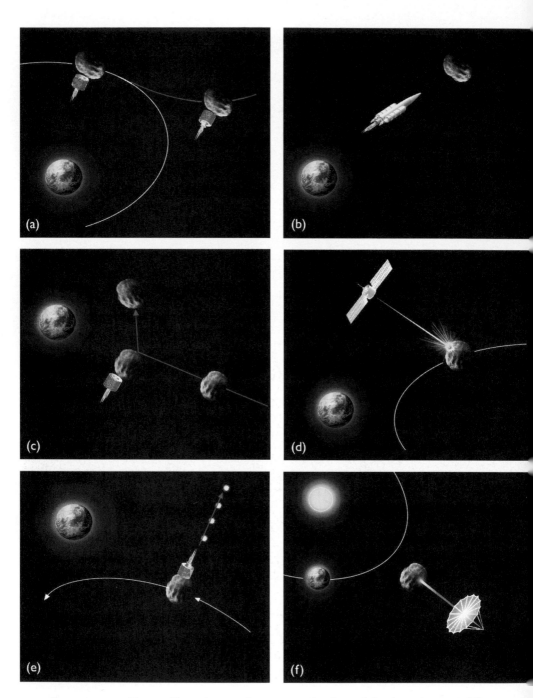

FIGURE 24: *Diagram illustrating several ways proposed to date to deflect incoming cosmic killers: (a) implanted rockets; (b) nuclear warhead blast; (c) collision with spacecraft; (d) laser or microwave blasting; (e) mass driver; (f) solar sailing (momentum from the Sun's radiation pushes it away, as wind pushes an umbrella).*

FIGURE 25: *An artist's rendition of the NEAR Shoemaker spacecraft as it orbits asteroid 433 Eros. (Courtesy of NASA.)*

FIGURE 26: *The eclipse of August 11, 1999: the diamond ring. (Photo by George T. Keene.)*

FIGURE 27: *Inca main ceremonial complex in Machu Picchu, Peru. (Photo by Michael J. P. Scott/Stone.)*

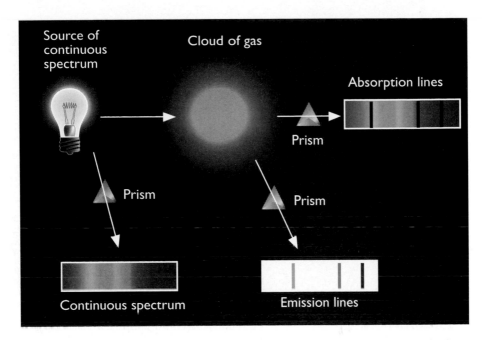

F I G U R E   2 8 .   *Emission and absorption spectra. A source producing a continuous spectrum (such as a light bulb) will have some of its lines absorbed as its radiation passes through a cloud of cool gas* (top sequence). *If, instead, one examines the spectrum from a single warm gas, it will reveal its very particular emission lines* (bottom right sequence).

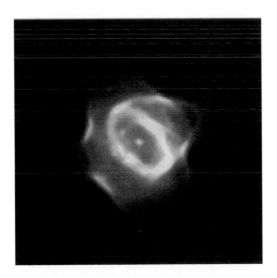

F I G U R E   2 9 :   *The Stingray nebula, the youngest known planetary nebula, as seen by the Hubble Space Telescope. (Courtesy Space Telescope Science Institute.)*

FIGURE 30: *In 1987, a supernova detonated in the Large Magellanic Cloud, a small satellite galaxy sometimes visible with the naked eye in the Southern Hemisphere. The progenitor star, being fifteen times heavier than the Sun, put on a spectacular display of the amazing power liberated in such events: at its peak, the supernova reached a luminosity of*

100 million Suns, outshining all the other stars in the Large Magellanic Cloud combined. The arrow points to the location of the progenitor star. (Copyright Anglo-Australian Observatory. Photo by David Malin.)

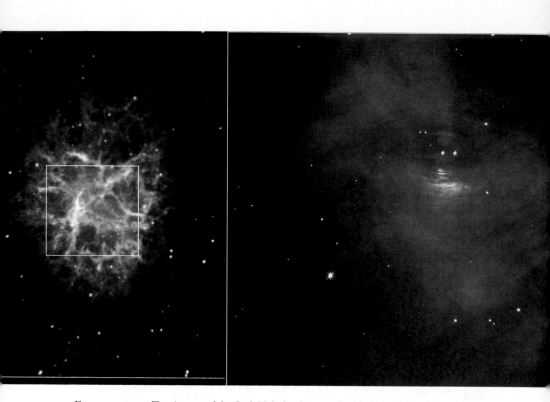

FIGURE 31: *Two images of the Crab Nebula, the most famous historical supernova, which flared in 1054. Left, an image obtained from an Earth-based telescope (the 5-meter Hale telescope) at Mount Palomar Observatory. Right, an image from the Hubble Space Telescope, magnifying the portion within the square of the left picture. There are two small stars at the center; the one to the left is a pulsar, a rapidly rotating neutron star. (Courtesy Space Telescope Science Institute.)*

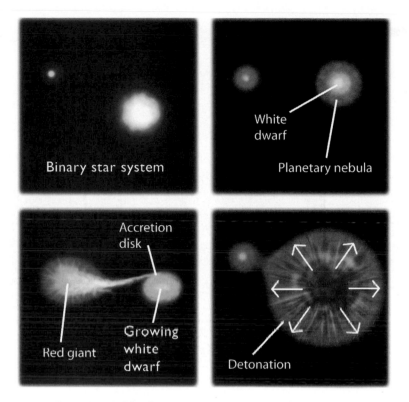

FIGURE 32: *Formation of a Type I supernova. A white dwarf in a binary system accretes matter from its companion until the degenerate electron pressure can no longer support it against collapse. The collapsing star fuses heavy elements at a furious pace until it detonates as a supernova.*

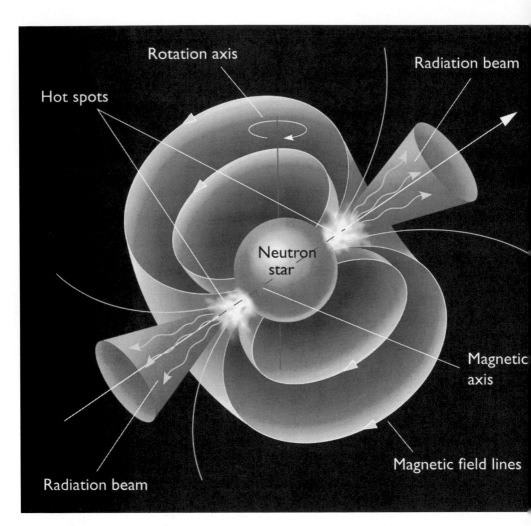

Rotation axis

Radiation beam

Hot spots

Neutron star

Magnetic axis

Magnetic field lines

Radiation beam

FIGURE 33: *The pulsar as a cosmic lighthouse. Beams of radiation are accelerated by the pulsar's magnetic field, which is not aligned with its rotation axis. As the beams sweep through Earth, we detect a very periodic signal.*

FIGURE 34: *Map by Olaus Magnus (sixteenth century) depicting the Maelstrom,* lower right. *(Reproduced from Giorgio de Santillana and Hertha von Dechend,* Hamlet's Mill *[Boston: Gambit, 1969], p. 91.)*

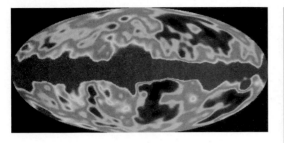

FIGURE 35: *Map of temperature fluctuations in the cosmic microwave background obtained by the COBE satellite. The fluctuations are only on the order of one hundred thousandth of a degree. (Courtesy NASA.)*

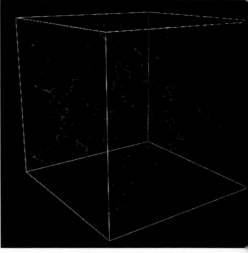

FIGURE 36: *Computer simulation showing the frot. nature of the large-scale structure of the universe. Reddi points denote larger concentrations of matter, such as ga ies and galaxy clusters. (Courtesy G. L. Bryan and M. L Norman, National Center for Supercomputer Applicatior*

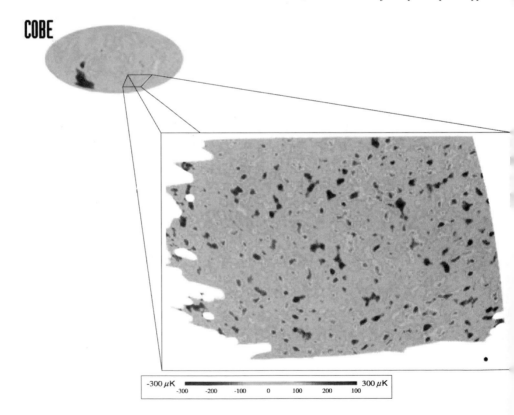

FIGURE 37: *Two maps of the cosmic microwave background fluctuations.* Top left, *the COBE map, with an accuracy of about ten degrees.* Bottom right, *the BOOMERANG map, with an accuracy of one degree. (COBE map courtesy NASA; BOOMERANG map courtesy BOOMERANG Collaboration.)*

sulfur, sodium, calcium, silicon, manganese, gold, and uranium come from? The answer is, perhaps, not much of a surprise: from stars heavier than the Sun. The fusion paroxysm that drove the Sun to its end will continue and produce heavier elements in more massive stars. In a very true sense, the chemistry of the universe is determined by stellar evolution. We should think of stars not as mere pale dots of light scattered around galaxies but as engines that give the cosmos its rich chemical diversity, our existence being just one of their many by-products.

Stars lighter than about eight solar masses will settle into white dwarfs, because the core temperatures will never rise above the 600 million degrees needed to fuse heavier elements. There is just not enough mass in the outer layers to compress the core to higher temperatures. To go back to the sleeper-on-the-mattress analogy, this sleeper is just too light to really squeeze the mattress down.

Stars heavier than eight solar masses will be hot enough at their core to continue where the Sun stops; instead of halting fusion at carbon (and a bit of oxygen), there will be a progressive buildup, based on a series of different fusion reactions. The most common involve the successive capture of helium 4 nuclei: carbon 12 captures helium to form oxygen 16 (at 200 million degrees); oxygen 16 captures helium to form neon 20; neon 20 captures helium to form magnesium 24; magnesium 24 captures helium to form silicon 28. If you guessed that there is a pattern here, you are right. Each newly fused nucleus has four more protons than its predecessor; two protons and two neutrons from helium 4, which get converted into protons. Because of the efficiency of these helium-capture reactions, "multiples of four" chemical elements are the most abundant in the universe. But clearly they are not the only ones. Other elements are fused as single protons and neutrons are freed from their parent nuclei and then absorbed by others.

This pattern of successive fusion by fours is mirrored in the very structure of the star; if stars like the Sun can be thought of as having an onion-like structure, with a carbon core surrounded by shells of helium and hydrogen burning (figure 13), here we will have a larger number of shells following the multiples-of-four hierarchy: from inside out, silicon 28, magnesium 24, neon 20, oxygen 16, carbon 12, helium 4, and hydrogen, as indicated in figure 14. When the star fuses silicon 28, another process starts to compete with helium capture, called photodisintegration, the breakdown of heavier nuclei by heat. The star's core is now at an absurd 3 billion degrees, and the radiation (gamma-ray photons) associated with this heat is so intense that it can break down heavier nuclei as easily as we can crumble a sugar cube. For example, silicon 28 photodisinte-

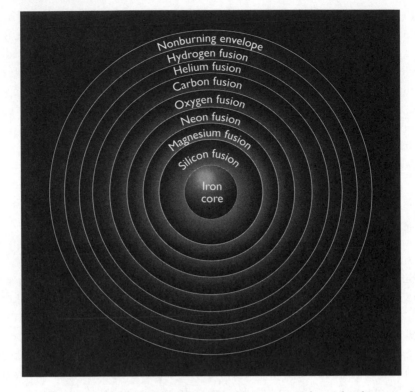

FIGURE 14: *A slice through a highly evolved star, with mass greater than eight solar masses. Note the onion structure, with heavier elements burning closer to the center at higher and higher temperatures. The star is about to detonate as a supernova.*

grates into seven helium 4 nuclei. This surplus of helium 4 is actually very good news, because it breathes new life into the helium-capture process, allowing heavier nuclei to get built up, all the way to nickel 56 (see figure 15). At this point, something quite different happens, which will radically alter the fusion-burning cycle of the star; nickel 56 is an unstable isotope of nickel (with 28 protons) and decays very quickly into cobalt 56 (with 27 protons). But cobalt 56 is also unstable, and it decays into the *very* stable iron 56 (with 26 protons). I say *very* stable because iron 56 has the most stable nucleus in nature. This means that it has the most tightly bound arrangement of protons and neutrons of any of the elements; iron 56 has the hardest nucleus to break apart. Not even the furious gamma-ray photons corresponding to temperatures of a few billion degrees can do anything with iron 56; it is as if the sugar cube turned rock-solid. The net result of this sequence of reactions is that the core quickly becomes iron rich.

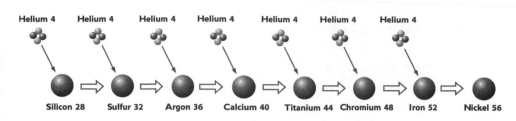

FIGURE 15: *Continuous fusion of helium 4 nuclei builds up heavier elements, all the way to nickel 56.*

This sequence of fusion steps is quite frantic; the heavier the element, the shorter the duration of the fusion process. As the star struggles against the crush of gravity, it burns whatever fuel it has available, heavier and heavier nuclei at higher and higher temperatures. In approximate numbers, a star twenty times as massive as the Sun burns hydrogen for ten million years (the heavier the star, the shorter its main sequence burning stage), helium for one million years, carbon for one thousand years, oxygen for a year, and silicon for a week. The iron core evolves within less than a day. Iron gets fused at the core not just from the decay of nickel 56 but also from the direct fusion of two silicon 28 nuclei, the closest layer to the iron core. On top of this, other unstable heavy nuclei also end up as iron. As more iron accumulates at the core, the star finds itself in a terrible predicament; being so stable, the fusion of iron 56 actually consumes energy instead of liberating it. The iron core stops providing the heat pressure that is badly needed to stop its own gravitational crunch; quite suddenly, the star loses its foundation and starts collapsing upon itself. As the core temperature rises to about 10 billion degrees Kelvin, the gamma-ray photons become so energetic that they can split even the iron nuclei. In fact, at these temperatures, photodisintegration runs wild and every nucleus at the core gets demolished into protons and neutrons; in fractions of a second, the photons undo what took millions of years to build through fusion. The once almighty star is being reduced to nuclear rubble from the inside out.

Paradoxically, photodisintegration costs energy; it's not easy to break down tightly bound nuclei into their constituents. This usage of energy cools off the core, further reducing its thermal pressure and opening the way for more gravitational compression by its outer layers; the star's heaviness becomes unbearable. Electron degeneracy pressure, so useful for lighter stars, is helpless here; as the squeezing mounts, protons and electrons are smashed together with such force that they morph into neutrons and neutrinos: proton + electron →

neutron + neutrino. This furious particle mutation at the core happens in less than one second! The impervious neutrinos carry away more energy, and the collapse proceeds at an even faster pace. The core is now a giant contracting ball of neutrons, which, because of their zero electric charge, can be squeezed much closer to each other than electrons can. If for white dwarfs the core densities are on the order of hundreds of thousands of grams per cubic centimeter (a few tons per cubic inch), for these neutron balls they can reach hundreds of *trillions* of grams per cubic centimeter (billions of tons per cubic inch); a mountain squeezed into a sugar cube! At these incredible densities, even the electrically inert neutrons react; they are squeezed so close together that, as with electrons for white dwarfs, their degeneracy pressure finally starts to counterbalance the central core's collapse.

This transition from indifference to degeneracy pressure is not smooth; the core's outer layers, still not converted, fall over the degenerate neutrons with a vengeance. Full of degeneracy pride, the neutron core acts as a brick wall, rebounding anything that hits it; the rebounding matter causes a huge shock wave that propagates outward at speeds of 50,000 kilometers per second, one-sixth the speed of light. Here two things can happen. One possibility is that the enormous violence of the shock wave literally rips the star apart, spewing its outer layers into space in what is known as a core-collapse supernova explosion, a Type II supernova, one of the most spectacular and energetic events in the universe. (Why Type II? You will find out very soon.) For a few days, a supernova may shine brighter than its host galaxy, with billions of stars (see figure 30 in insert). The enormous release of energy through its outer layers cooks elements heavier than iron; the heavier chemistry of the universe is the signature of the dying star's last convulsions. Another possibility is that the shock wave reverses itself on its way out and that the outer layers of the star come back crashing down on the core. In this case, not even the neutrons can hold the star back, and the core collapses into a black hole, an object with such intense gravity that not even light can escape its gravitational grip. But this is not a tale only of decay and destruction. Five billion years ago, a nearby supernova quite possibly triggered the collapse of the hydrogen cloud that became our solar system. The explosion sprinkled our progenitor cloud with the chemical elements needed for life to develop on Earth. Explosions destroy and explosions create.

The blast from a supernova may leave behind a highly compacted neutron ball, which is appropriately called a neutron star, a star of nuclear matter, with a mass comparable to the Sun's but with a radius of only about 10 kilometers. Bizarre as they are, neutron stars still strike me as the last bastions of normalcy, where physical laws, even if in extreme circumstances, can be applied and meaning be extracted from them. The situation changes with black holes. Of course, astrophysicists have studied the properties of black holes in great detail. But in spite of all their success, or because of it, the study of black holes has also raised many questions, which, as we will see, remain very much unresolved. Before we plunge into a more detailed study of the properties of neutron stars and black holes, let us briefly visit the history of supernovae explosions, and understand why there are two types of possible supernova blasts, Types I and II.

New Stars That Are Old

Supernovae have baffled sky watchers for millennia. Chinese records from 185 c.e. tell of a new star in the constellation Centaurus that shone for twenty months, disappearing as mysteriously as it appeared. Another was recorded on 393 c.e. Most certainly, the court astrologers viewed these events with much trepidation, as celestial portents of hard times to come. Modern astronomers have identified the remnants of these explosions, wisps of gas rapidly expanding through the interstellar medium. In 1006, the skies once again featured a new light; this time, not just Chinese but also Arabic, European, and Japanese sky gazers recorded the unusual event, a new star that flared in the constellation Lupus and was so bright as to be visible during the day for months and at night for three years, initially even rivaling the Moon.

The most famous of all supernovae flared in 1054 in the constellation of Taurus. An entry in the records of the Imperial Observatory of Peking reads,

> In the first year of the period Chihha, the fifth moon, the day Chi-chou [July 4, 1054], a great star appeared approximately several inches southeast of T'ien-Kuan [i.e., the star Zeta Tauri]. After more than a year it gradually became invisible.[2]

This supernova flared during the day for three weeks, and at night for two years. A cave painting in Chaco Canyon, New Mexico, attributed to the Anasazi, shows quite clearly the bright star close to the waning crescent Moon, precisely as it appeared in the skies on July 5, shining brighter than Venus. The stellar explosion left a beautiful expanding remnant known as the Crab Nebula, first cataloged by the French astronomer Charles Messier in the 1770s under the symbol M1—the first object of the Messier catalog of celestial objects. Little did he know that the core of the nebula nested another remnant of the explosion, a rapidly rotating neutron star discovered only in 1968, spinning at an amazing thirty revolutions per second (see figure 31 in insert). Imagine an object made of neutrons, not much larger than Mount Everest, with a mass comparable to the Sun's, and spinning around its axis thirty times a second! How can such an object exist?

Two other historical supernovae were extremely important to the development of Western science, those of 1572 and 1604. We encountered the supernova of 1572 in chapter 3, the "new star" in the constellation of Cassiopeia seen by Tycho Brahe on the evening of November 11, as he walked back from his alchemy laboratory. Although the supernova was seen by many other European astronomers, it was Tycho's detailed quantitative analysis that conclusively demonstrated the star to be farther away than the Moon, thus delivering a severe blow to the Aristotelian view of an unchangeable cosmos. Johannes Kepler, the visionary German astronomer who first formulated the laws of planetary motion, recorded with great excitement the supernova of 1604, which appeared conspicuously close to a near-conjuction of Mars and Jupiter; all three celestial objects were located within a very small region of the sky, a sign charged with the gloomiest astrological significance, which Kepler and many others were quick to point out. In spite of his strong spiritual (and lucrative) attraction to astrology, Kepler was very much a man of a transitional era, when accuracy and precision of astronomical measurement was first being combined with serious scientific questioning regarding the physical causes of natural phenomena. As his former employer Tycho had done before, Kepler emphasized the great distance to the new celestial luminary, which, unlike planets, did not move with respect to other stars. It is an interesting irony that, although supernovae have been historically associated with new stars, they actually mark the end of a star's existence, being "new stars" that are really old and dying. The "new star" misnomer is a consequence of observational limitations; with the naked eye or small telescopes, it was (and even today still is) hard to realize that the "new star" was actually there all along, just not as obviously visible as others because of its distance (and thus faintness) or of obscuration by interstellar dust. Dating back to the first recorded supernovae, almost two thousand years of astronomical research were necessary for scientists to reach this conclusion.

The supernovae described above, created when the shock wave from the rebounding core rips apart the outer layers of the agonizing star, are not the only kind possible. This explains the "Type II" title I used earlier, implying that there must be a Type I. For Type I supernovae, the enormous explosion actually takes two players, two stars that travel around each other in a kind of cosmic merry-go-round, bound together by their mutual gravitational attraction. These systems of two stars are called binary systems and actually make up most of the stars in our galaxy; our Sun is an atypical loner.

Let us, then, imagine a binary system where one of the stars became a white dwarf. If the two stars are close together, the white dwarf's gravitational pull on its companion will be so intense that it will be able to suck the matter in its outer layers, like a cosmic drain (see figure 32 in insert). This process, known as *mass accretion,* is likely to occur if the companion star is also near its death, say, within its red giant phase. The white dwarf will accumulate hydrogen and helium in its outer layer, which may ignite explosively into thermonuclear fusion, ejecting huge amounts of hydrogen into space. These explosions, less powerful than supernovae, are known as novae. They don't disrupt the stability of the white dwarf and can actually recur several times. To actually turn a nova into a supernova, the white dwarf must accrete enough matter for its mass to grow above the critical value of 1.4 solar masses, known as the Chandrasekhar limit. Above this mass, the electron degeneracy pressure cannot support the star any longer, and the gravitational squeezing causes the pressure and temperature at the core to soar, reawakening the fusion chain. Carbon fuses into heavier elements at a frantic pace everywhere within the white dwarf, culminating with an enormous explosion of power comparable to (and sometimes even greater than) the supernovae produced by the collapse of heavy stars. To distinguish between Type I and Type II supernovae, astronomers examine the spectrum emitted by the exploding objects. Since a Type I starts with carbon fusing into heavier elements, its spectrum will be very poor in hydrogen. The opposite is true of Type II supernovae, whose hydrogen-rich outer layer is duly recorded in their spectra.

An Absurd Idea

The attentive reader must have noticed that I mentioned a critical mass for white dwarfs of 1.4 solar masses without explaining where this odd number comes from. The discovery of the stability limit of white dwarfs has a very interesting story, well worthwhile retelling here. During the 1920s and 1930s, the astronomy scene in England was dominated by the towering figure of Sir Arthur S. Eddington. In his 1925 book *The Internal Constitution of Stars,* Eddington raised the issue of the stability of white dwarfs, which, to him, was a great mystery. Like all astrophysicists in the early 1920s, he believed that the stability of a star, any star, despite the gravitational squeezing of its outer layers could be due only to the pressure caused by the very hot atoms in its central region. If a star cools down, its heat pressure will decrease and the gravitational squeeze will increase. As this happens, the star gets smaller and matter at the core gets hotter again, creating more heat pressure until a new state of balance is achieved. Then the star starts to cool, and the whole process is repeated. This sequential shrinking truly puzzled Eddington; if, for each new state of equilibrium, the cooling star is smaller, will there come a point when the star shrinks into "nothing" and disappears? Eddington refused to accept this radical idea (actually similar to what happens when a black hole forms, as we will soon see) and offered a (wrong) way out, conjecturing that the only other counterpressure to balance the star is the electric repulsion of the squeezed adjacent atoms, the same that gives rocks their solidity.

The problem with Eddington's suggestion is that it required the stellar matter to behave like rocks, that is, to have rocklike densities of a few grams per cubic centimeter, which was tens of thousands of times less dense than white dwarfs were then said to be. Eddington expressed his indignation with these huge densities in strong wording: "I think it has generally been considered proper [when discussing ultra-high-density stars] to add the conclusion 'which is absurd.'"[3] At the time, astronomers knew that the brightest star in the night sky, Sirius, actually has a very small and dense companion, known as Sirius B. Their measurements indicated that Sirius B has a mass equivalent to 0.85 solar masses and a radius of 18,780 kilometers. The numbers today are 1.1 solar masses and a radius of 5,500 kilometers, giving Sirius B a mean density of three million grams per cubic centimeter, a value even more "absurd" than that prevalent in Eddington's day. He conceded that the white dwarfs were "a curious problem and one may make fanciful suggestions as to what actually will happen."[4] And fanciful they were indeed.

The first step toward the resolution of Eddington's white dwarf paradox was put forward by the British astrophysicist R. H. Fowler in 1926, in an article titled "On Dense Matter." Using a brand-new development in quantum mechanics, Pauli's exclusion principle (see chapter 5), Fowler brilliantly pointed out that Eddington's argument for the stability of white dwarfs on the basis of classical physics was flawed; at the enormous densities at the cores of white dwarfs, electron degeneracy pressure, and not heat, provides the required support. Fowler concluded that the physics of the very small had a crucial role in the physics of compact (high density + small size = compactness) astronomical objects, changing forever the course of astrophysical research.

Far away in Madras, India, a teenager named Subrahmanyan Chandrasekhar, still working toward his bachelor's degree, fell in love with this new astrophysics. These were also the days when Einstein's general theory of relativity, a radically new way of understanding gravity, was receiving great acclaim, some of its predictions being confirmed by none other than Eddington himself. What could be more exciting for an emerging physicist than to work on a topic that mixed the two great new theories of the time?

Chandrasekhar's dream came true in 1930 when, at the age of nineteen, he sailed to England to pursue graduate studies at Cambridge University, the home of his idols, Fowler and Eddington. He did not waste any time. During the eighteen-day trip, he decided to expand on Fowler's work; there were lots of open questions, most notably how the degenerate electrons would respond to an increase in the squeezing caused by a larger mass. Would their degeneracy pressure increase so as to counterbalance the mounting gravitational crunch? Was there a limit to how much gravitational squeezing the electrons could take? The answers are yes and yes. As the outer layers of the white dwarf squeeze the degenerate electrons, they will respond by moving faster and faster in their "exclusive" cells, kicking indignantly against the imposed loss of space. Chandrasekhar realized that at densities comparable to that of Sirius B, their velocities were about 57 percent the speed of light. Einstein had demonstrated in 1905 that no massive body could reach the speed of light, electrons included. In his special theory of relativity, he predicted that any body that is accelerated to speeds close to the speed of light will respond by increasing its mass; that is, fast motion makes things heavier. At 57 percent the speed of light, some changes, even if small, are to be expected.

The fact that we don't notice these effects, because of the very low speeds of everyday life—much smaller than 300,000 kilometers per second—does not mean they are not observed in extreme situations, for example, in experiments involving highly energetic particle collisions. In the interior of white dwarfs, the extra energy due to all the squeezing is partially transformed into mass and not into larger speeds, with catastrophic consequences for the star's stability. Even though at the time there was no theory merging quantum mechanics with special relativity, Chandrasekhar was able to demonstrate by means of approximate arguments that white dwarfs could not be stable if their masses were above 1.4 solar masses; the extra energy spent making the electrons "heavier" compromised their ability to counteract the gravitational squeezing, causing the star to implode. So, at nineteen, Chandrasekhar predicted that white dwarfs could not have masses exceeding 1.4 solar masses, a truly remarkable result.

But Chandrasekhar had an uphill battle to fight. He summarized his results in two papers, one about slow degenerate electrons and the other about fast electrons and his predicted mass limit, and sent them to Fowler for publication. Fowler published the first one, about normal white dwarfs, but was suspicious of Chandrasekhar's results for relativistic electrons. After a few months, Chandrasekhar decided to submit his mass limit calculations to the American publication *The Astrophysical Journal*. The paper was accepted, but, quite to his shock, nothing else happened after its publication; the astronomical community decided to ignore his results on the critical mass for white dwarfs. Since he had to finish his Ph.D., for the next three years Chandrasekhar decided to take on less controversial subjects. Doctorate in hand, he resumed the attack. Inspired by some colleagues, he decided to check the masses of ten known white dwarfs to prove that they were all below his critical mass limit. It was not definite proof, but certainly strong circumstantial evidence that his theory was consistent with observation. After many days of hard work, which included using a computer—a 1934 computer—to solve the complicated equations derived from his theory, Chandrasekhar got what he wanted; all ten stars had masses below the 1.4 solar mass limit, the masses behaving in excellent agreement with his theory. He was now convinced that the astronomical community would have to accept his results.

Not so easy! Chandrasekhar was openly confronted by Eddington in a charged night at the Royal Astronomical Society of London, where, to his surprise, they both were scheduled to give presentations; although Eddington had been following Chandrasekhar's progress closely (even the computer was his), he never mentioned his own work on the subject. After Chandrasekhar presented his results, it was Eddington's turn. He criticized Chandrasekhar's work, declaring that "there should be a law of Nature to prevent a star from behaving in this absurd way!"[5] It was very hard for him to accept the implications of Chandrasekhar's results, that collapsing stars could lead to a breakdown of the laws of physics. His point—which was wrong—was that Chandrasekhar's meshing of relativity and quantum mechanics was incorrect and thus so were his conclusions. Eddington proposed his own combination of the two theories, with the result, not surprisingly, that white dwarfs could be as massive as they had to be; they were the corpses of any mass star. Chandrasekhar instinctively knew that Eddingon's ideas were wrong; but what could he do, when the greatest astrophysicist of the time challenged a recent Ph.D.? The astronomical community, without better guidance, listened to authority, something that happens more often in science than it is thought. Controversy is part of everyday life for scientists, since skepticism is the only attitude that protects science against fraud and charlatanism. However, many controversies, in the absence of enough data or observational evidence, are settled on the basis of authority, at least at first. What saves scientists from stubbornly supporting an erroneous theory is the continuous scrutiny to which ideas, even those accepted as correct, are submitted. Sooner or later, an idea initially dismissed as wrong may emerge as the correct explanation, even though this may sometimes take decades. This was the case with Chandrasekhar's mass limit. Chandrasekhar was so burned out by the whole episode that he did not work on white dwarfs for thirty years. In 1982, he was awarded the Nobel Prize in physics for his remarkable contributions to astrophysics, and in particular for his prediction of a critical mass for white dwarfs. Scientists, being only human, make mistakes. Fortunately, we have our colleagues to set us straight, a process that is as necessary as it is painful.

The Last Bastions of Normalcy

The implications of Chandrasekhar's mass limit were clear; if white dwarfs cannot have masses above 1.4 solar masses, what happens when the stellar core is more massive than that? There are countless stars much more massive than the Sun, up to ten or twenty times more massive. Surely their collapse and supernova detonation would leave a core heavier than 1.4 solar masses. We have seen that another state of matter is expected to emerge as the core pressure mounts beyond the white dwarf limit, when protons and electrons are squeezed together into neutrons at densities of hundreds of trillions of grams per cubic centimeter. These neutron stars were first conjectured by the controversial (and brilliant) astronomer Fritz Zwicky in collaboration with Walter Baade, both working at Caltech in the mid-thirties, the time of the Chandrasekhar-Eddington confrontation. Similar ideas were also proposed by the great Russian physicist Lev Landau, even before Zwicky; it is said that Landau thought of the possibility of neutron stars—or at least neutron cores in stars—the same day that Chadwick announced the discovery of the neutron in 1932. But Landau did not publish his results until 1937, and then as a desperate attempt to help him escape the widespread purges underway in Stalin's regime. Unfortunately, his plan did not work; he was imprisoned the next year, and subjected to horrible treatment until his release in 1939. Although Landau did recover from his year in prison, to the point of doing pioneer work in low-temperature physics and receiving a Nobel Prize for it, he was never the same again; if Stalin's KGB failed to destroy his genius, it surely destroyed his spirit.

In a classic paper from 1934, Zwicky and Baade proposed that supernovae are astronomical objects different from novae, and that they flare during a star's transformation into a neutron star. They even coined the name *neutron star.* However prescient Zwicky and Baade were, they left the crucial question unanswered: If white dwarfs have a mass limit, what about neutron stars? And if neutron stars have a mass limit, what happens next? The transition from degenerate electrons to degenerate neutrons was far from trivial. Whereas electrons interact by their electric repulsion, neutrons interact by the strong nuclear force, which posed several theoretical difficulties at the time and, to a certain extent, even today. Nevertheless, two other Caltech physicists had enough knowledge of nuclear physics to develop, at least approximately, the basic theory behind neutron stars: Richard Chace Tolman and J. Robert Oppenheimer. The latter was to become the well-known leader of the Manhattan Project, which built the American atomic bomb a few years later.

As was his style, Oppenheimer enlisted a student, George Volkoff, to work out the details. His approach was quite logical; since they did not know much about how the strong nuclear force acts on ultradense neutrons, they should first obtain the critical mass limit for neutron stars assuming that there is no strong force at all. They found that, in the absence of strong nuclear forces, neutron stars do indeed have a critical mass, and only of 0.7 solar masses. This was quite an unexpected result. After all, if neutron stars have a mass limit, the idea of black holes had to be taken quite seriously; a star that had a neutron core heavier than the critical limit would keep on collapsing, with nothing to stop it from its terminal implosion. In came Tolman, with his estimates of how neutrons should behave if the strong nuclear force between them was either highly attractive or repulsive. Once these two hypothetical limits were factored in, the original estimate changed. But not by much. The trio concluded—correctly, we now know with confidence—that neutron stars do have a critical mass between half and a few solar masses. After fifty years of observations, and with much more sophisticated knowledge of the strong nuclear force, astrophysicists have concluded that neutron stars do have a limit between 1.5 and 3 solar masses, not far from the Oppenheimer-Tolman-Volkoff prediction. What we still do not understand is why the observed neutron stars, now in the hundreds, all seem to have around 1.4 solar masses.

It was only in 1967 that neutron stars were first observed. Jocelyn Bell (now Bell Burnell) was working toward her Ph.D. at Cambridge University, under the supervision of Anthony Hewish. Her research consisted of developing and testing radio telescopes designed to measure emissions from distant radio sources. Just as the Sun radiates strongly in the visible and infrared parts of the electromagnetic spectrum, different astronomical sources may radiate mainly at different frequencies, ranging from the low-frequency (and low-energy) radio waves to the high-frequency (and high-energy) gamma rays. Thus, we may "see" sources that are actually invisible to our eyes through specially designed telescopes, sensitive to different kinds of radiation.

Jocelyn's task was extremely tedious—to manually check hundreds of meters of tape with the recorded data from the radio telescope, which gave the intensity of the received radiation as time went by. Research is not always—in fact, seldom—glamorous, often involving detailed analysis of huge amounts of data, or fixing mistakes in computer programs thousands of lines long, or plowing through pages and pages of calculations. But it is all (almost always) worthwhile in the end, when results are finally obtained—especially results such as those obtained by Jocelyn. She found an intriguing signal in her data, a peak in the intensity of received radiation, that was repeated with incredible accuracy every 1.33730133 seconds, somewhat like a very steady heartbeat in an electrocardiogram. The radio source was emitting pulses of radiation as regularly as the ticktack of a clock! Not knowing what these were, Bell Burnell and Hewish decided to name the object LGM, for Little Green Men; after all, astronomical sources were believed to be large, generating radiation of low frequency, typical of slowly moving matter. The possibility that a source could be at once so regular and fast triggered the imagination of the astronomical community and the popular press, and the Little Green Men hypothesis was taken quite seriously by a number of people. Maybe the signals were indeed produced by a distant intelligent civilization, eager to establish contact with other intelligences in the universe.

Soon after this shocking discovery, Bell Burnell and Hewish found other rapid sources of radio pulses, with different periods, which became known as *pulsars;* so much for the Little Green Men hypothesis. The theorists Franco Pacini and Thomas Gold suggested that these amazing pulse sources were rapidly rotating neutron stars; as planets rotate about their axis, so do stars and pretty much everything else in the universe. If you recall our discussion of angular momentum in chapter 3, as a spinning chunk of matter contracts, it spins faster. This will also be the case for a rotating massive star undergoing gravitational collapse; if the star rotates about its axis once every couple of weeks, as heavier main-sequence stars are observed to do, by the time its core shrinks into a neutron core 20 kilometers in diameter it will rotate a few times per second or faster. No earthly matter could withstand the enormous centrifugal forces from this mad rotation; but ultradense neutrons can, at least up to periods of about a thousandth of a second, or a millisecond. At those incredible rotational velocities not even degenerate neutrons can stay together.

Fast rotation is a natural outcome of the way neutron stars are formed. But what about the radio pulses? They are generated by matter being swirled around in the highly focused and intense magnetic fields that are also produced during the formation of neutron stars. Ordinary stars, along with many planets, including Earth, have large-scale magnetic fields. Just as the star's mass density at the core increases with its collapse (same amount of mass in a smaller volume implies larger density), so does its magnetic field; as the star collapses, its magnetic field is also squeezed into a smaller volume, reaching enormous values. Contraction amplifies not only the gravitational pull of a star but also its magnetic field; we may think of contraction as a focusing mechanism for the magnetic field, just as when we focus light under a microscope, bringing the rays "closer together" for a better image. The resulting fields are amazing; the typical magnetic field of a neutron star is trillions of times larger than Earth's field. A person wearing a belt with a steel buckle close to a pulsar would be flung away at supersonic speeds. (If he got too close to the star, he would also be flattened thinner than a sheet of paper by the huge gravitational pull at a neutron star's surface. Not a very hospitable environment.)

More often than not, the magnetic field is not aligned with the rotation axis of the star. An example of a misalignment between two "axes" familiar to us is a lighthouse, where the highly focused beam of light makes a right angle with the rotation axis, which points up. In fact, the lighthouse analogy is extremely useful when picturing a pulsar; substitute for the beam of light the magnetic field axis, and you have a rapidly rotating magnetic field, although for pulsars the angles are smaller than 90 degrees (see figure 33 in insert). Electrically charged particles left over from the parent star are accelerated to very high energies by the rotating magnetic field, emitting radiation within a narrow cone aligned with the magnetic field axis, just like light coming out from a lighthouse. If we happen to be at the right spot, this radiation will pass by us as the magnetic axis sweeps across the cosmos. And if a radio telescope (or other telescopes, like x-ray, gamma-ray, and even sometimes optical) points in the direction of the beam, it will detect regular pulses every time the beam sweeps its surface. This mechanism correctly explains the mystery of the Little Green Men; some pulsars are so incredibly accurate that their period is predicted to change by only a few seconds over millions of years, which makes them far more accurate than the best atomic clocks on Earth.

There have been only a few direct associations between pulsars and supernova remnants, most notably the one that flared in the Crab Nebula in 1054, six thousand light-years from Earth (see figure 31 in insert). Astronomers using both radio and optical telescopes identified a very fast pulsar there, spinning thirty-three times per second, faster than our eyes can detect. The same way that the mass determines the fate of a star, the magnitude of the magnetic field determines the fate of a pulsar; the pulsar slows down as it dissipates energy through its rotation, and it practically stops after a few tens of millions of years. Thus, the fastest pulsars are also the youngest, as is the case of the Crab pulsar, not even one thousand years old. The relatively short life span of pulsars may explain the lack of more matches between the hundreds of known supernova remnants and the hundreds of known pulsars; the pulsars are already "extinguished," turned into practically motionless neutron stars. Or they were never formed during the violent supernova explosion, which either tore the neutron core apart or created a black hole instead. Another explanation is that the pulsars may very well be there, but beaming at angles that simply miss us. Finally, supernova explosions may not be perfectly spherical; if there is a small asymmetry in the core collapse, the neutron star may recoil from the center of the explosion at speeds of several hundred kilometers per second. The supernova remnant and the pulsar would then be at odd locations, making it hard for astronomers to match the two.

There are many things we still do not understand about neutron stars and pulsars. Neutron matter at the core of a neutron star may assume forms that are at the edge of our present understanding of high-energy particle physics. There have been several extremely interesting proposals to model matter at densities of thousands of trillions of grams per cubic centimeter, but they remain speculative; it is very hard to test physics at these densities! In any case, the question of most interest to us, whether neutron stars have a critical mass, was answered approximately by Tolman, Oppenheimer, and Volkoff just before the Second World War and with much higher precision over the past two decades; neutron stars do have a critical mass somewhere between 1.5 and 3 solar masses. That means more massive cores will continue their collapse indefinitely. As the core contracts further, gravity will grow progressively stronger, in a runaway process that will create the most fascinating object in the universe, a black hole.

Gravity Redefined

In order to appreciate fully the remarkable properties of black holes, we must explore the physics that dictates their behavior. Before we plunge into our own journey of exploration, we need to spend some time preparing for the trip. The hurried reader may skip to the last paragraph, where I summarize, very briefly, the main lessons that must be learned. However, I suggest that you all take a deep breath and stick with me for the next few pages, for the price is well worth paying.

Late in 1915, after struggling with confusion, exciting results, and false starts on and off for eight years, Albert Einstein, then only thirty-six years old, published his work on a new theory of gravity known as the general theory of relativity.[6] The theory marked a sharp departure from the then universally accepted explanation of gravitational attraction, proposed by Newton in 1687. In the Newtonian theory, the gravitational attraction between two bodies acted "at a distance," through a force proportional to the product of their masses and inversely proportional to the square of their distance. Newton wasn't clear about how two bodies could interact without touching, or what mysterious mechanism within the bodies generated the attraction; he preferred to "feign no hypotheses," being quite content with having developed a framework that could describe quantitatively the motions of celestial and terrestrial objects under the influence of gravity.

I would like to pause for a second and make an important remark about the structure of physical theories. Many people, inside and outside the classroom, have expressed to me their suspicions about Newton's "explanation" of gravity. "Really," they say, "how can physicists claim they *understand* gravity if they don't even know why massive bodies attract each other?" Well, we may not understand *why,* but we understand *how.* And to build a description of the physical world it is not critical to understand the whys of natural phenomena; the how is good enough. It is important to keep in mind that scientific explanations are built to *describe* the phenomena we observe in nature, after we make certain assumptions about what it is we want to describe. Like all that we create, science has its limitations, working quite well within its set boundaries. This does not mean that questions such as the origin of the gravitational attraction of masses are outside the scientific discourse; maybe one day we will understand it. But it does mean that while we don't, we can still make plenty of progress understanding how gravity shapes the natural world.

Newton's theory was built upon a rigid definition of space and time; in order to describe physical phenomena, he assumed that space and time are absolute entities, indifferent to the presence of an observer. This means that we can picture absolute space as the arena where things happen, a sort of stage for natural phenomena; as we observe them unfolding, the stage remains the same. More specifically, it means that people moving at different velocities measure the same distances between two points—say, the length of a stick—irrespective of their relative velocity. (For now, we limit the discussion to constant velocities.) "Not so," Einstein would have said. The reason why we think the two measurements give the same result is that our motions are very slow compared with the speed of light. Consider the following situation: one person is in a train moving at constant velocity with respect to another person standing at a station. Inside the train, there is a 1-meter-long stick, as measured by the person in the train. What would the person at the station measure? A *shorter* stick! The faster the train moves, the shorter the stick will appear to be, especially for speeds close to the speed of light in empty space, 300,000 kilometers per second.

Thus, according to Einstein, and this he knew already in 1905, observers in relative motion will measure different lengths; space is not as rigid as Newton thought.

The same holds for the passage of time. If the two observers in relative motion are measuring the time interval for a particular event, they will also get different results. As an illustration, consider again two observers, one at a train station and the other traveling in the train at some constant speed close to the speed of light. (This is an imaginary train, of course!) This very fancy train has a glass ceiling and is equipped with two large flashlights fastened to its top at opposite ends so as to face each other, as in figure 16. The glass ceiling allows the passenger inside to see what is happening at the top; she sits exactly at their midpoint. The flashlights were designed to flash *together* once every so often.

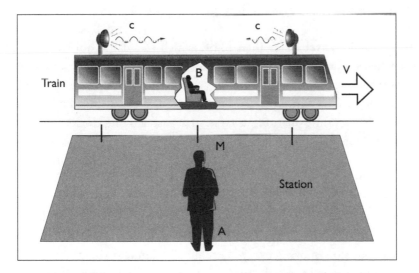

FIGURE 16: *Two observers, one at a train station (A) and one traveling in a train (B) moving at constant velocity, measure the time it takes for the light from two flashlights to reach them. Whereas for the observer at the station the two flashlights flash simultaneously, the observer in the train sees the front light pulse before the one from the back.*

Here is the experiment: as shown in the figure, as the train approaches the station, the two flashlights flash together precisely when the observer on the platform is at their midpoint. He makes a note of how long it took for the two flashes to hit him, each coming from one side, and he concludes they took the same exact time; for him, the flashing was simultaneous. "Not so," says the passenger in the train. "I also measured the time it took for the two flashes to hit me, and conclude that the flash from the front took a shorter time to hit me." Now, this is quite unexpected; two events that are simultaneous for one observer are not so for another!

This experiment has very broad implications; we describe natural phenomena by measuring how they behave in space and time, that is, where a certain object is when. If measurements depend on the relative motion of observers, unless we establish a set of rules so that they can compare their results, we wouldn't be able to make head or tail of the natural world, at least for things moving close to the speed of light. This "dictionary" is precisely what Einstein provided in 1905, with his special theory of relativity. His conclusions are a consequence of assuming that the speed of light is the fastest speed possible—the fastest speed at which information can travel—and that it is always the same, even if it is emitted by a moving source. This is where light differs from other objects; if you are in a car at 60 miles per hour and throw a ball at 20 miles per hour in the same direction you are moving, the ball's speed will be 80 miles per hour with respect to the ground. But not light; it will always move at the same speed!

If one could travel faster than light, which one can't, our nicely ordered reality where causes precede effects would go up in smoke; we could, in principle, travel backward in time, toward the past. The bizarre effects of length contraction and loss of simultaneity are consequences of this limiting speed. Another effect is "time dilation," the fact that fast-moving clocks tick slower. If the passenger in the train had a clock with her, the observer on the platform would "see" it ticktock slower. "But how do you know this really happens?" you may justifiably ask. Because the predictions from special relativity have been confirmed over and over again in laboratory experiments and in observations involving fast-moving elementary particles, where the speed of light is not such an enormous speed. Our perceptions of the natural world, immersed as they are in our slow-moving everyday reality, are quite myopic.

Revolutionary as the predictions of special relativity were, Einstein knew they were just the tip of the iceberg. After all, most motions we observe in nature do not happen at constant speeds but involve changing speeds, that is, acceleration. Somehow, a complete theory of relativity should be able to incorporate accelerated motion. At this point enters what Einstein dubbed "the happiest thought" of his life, which occurred to him as early as 1907. Lost in contemplation one day, he asked himself an apparently innocent question, which I paraphrase: "Imagine a person falling from a huge height. As she is falling, she feels weightless; falling eliminates the force of gravity!" Let me elaborate on this image through a more concrete example, a trip down a fast elevator in a high-rise. You get in on the top floor, squeezed among lawyers, stockbrokers, and office staff. Feet planted on the floor, you feel your full weight, as Earth pulls you down. (And you pull Earth up, but she doesn't seem to care much.) As the elevator starts its descent, you feel lighter, a queasy feeling in your stomach; most people are familiar with this experience. Suddenly, you hear a loud noise; the cables snap, and the crowded box falls vertically down, accelerating at a constant rate. Amid the screaming, you notice that your feet, and that of your fellow passengers, are not touching the ground any longer; you are free-falling with the elevator and feel weightless, very much like astronauts in a spaceship. Gravity, or at least the feel of it, is gone. (Of course, gravity is what is making you and the elevator fall at the same rate.) Fortunately, when the elevator reaches the tenth floor, the emergency breaking system kicks in; everyone hits the ground with a vengeance, feeling much heavier for a few floors, until the elevator slows down to its comfortable constant speed by the second floor. Your normal weight is back and all ends well.

This dramatic experience illustrates how acceleration can mimic the effects of gravity. If the elevator is not moving (or moving at a constant speed), Earth pulls you down and the floor pulls you "up" by the same amount and you feel your normal weight. As the elevator begins its descent, it must accelerate from rest, that is, with respect to Earth; this results in a decrease of the net force that you (your feet) feel and thus a decrease in your perceived weight. The larger the downward acceleration of the elevator, the lighter you feel; it is as if your feet could not keep up with the elevator's floor. When the elevator free-falls, the "up" force effectively disappears, and you feel weightless, free-falling with the elevator. The situation is reversed when the brakes start to slow down the elevator; the floor pushes hard against your feet, and you feel heavier.

Reasoning along similar lines, Einstein concluded that it is not possible a priori to truly distinguish between accelerated motion and the acceleration caused by gravity; inside the elevator, the changes in acceleration feel just like changes in weight, which is how your body (your total mass) responds to Earth's gravity. Had you not known that you were in an elevator on Earth, you could have argued that some mad scientist was toiling with Earth's gravity, making it weaker and stronger at will. The fact that accelerated motion can mimic gravity is known as the *principle of equivalence*. Thus, a general theory of relativity that incorporates accelerated motion must also be a theory of gravity.

By the time Einstein arrived at the final formulation of his theory, it had taken a surprising turn; he realized that there was a new way of thinking about gravity, quite different from the action at a distance that characterized Newton's theory. To Einstein, the gravitational pull a massive body exerts on others could be explained as a curvature of space around the body; the more massive the body, the more accentuated the curvature around it and thus the stronger its gravitational pull. This can be illustrated by a two-dimensional analogy, which, although limited, has its merits. Imagine a perfectly flat mattress. If you throw small marbles on it, they will move in straight lines at constant speeds (see figure 17). Now place a large lead ball at the center of the mattress; the surface of the mattress will curve around the ball, the curvature being larger the closer to the ball. If you throw a marble close to the ball, its path will no longer be straight, for it will move in a curved geometry; thus, the motion of the marble is influenced by the distortion the lead ball causes on the mattress. Furthermore, this distortion will cause the marble to accelerate. Einstein equated this acceleration with the distortion of the geometry, pointing out that, for small enough masses, even as large as planets, the motions would closely match the predictions of Newtonian gravity. In other words, the effects of curvature predicted by general relativity will be relevant only for very large masses, such as stars. The rigid space of Newtonian physics was gone for good; Einstein added plasticity to space, which deformed in response to large masses placed in it.

But space was not the only one of the two Newtonian absolutes that general

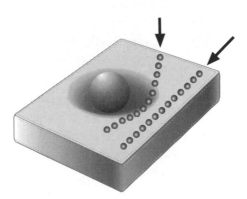

FIGURE 17: *Illustration depicting a mattress, which is deformed by a heavy ball placed on its surface. Notice how the trajectories of two marbles that pass by the ball are different, depending on their proximity to the ball.*

relativity revised. The flow of time will also be modified by the presence of large massive bodies—the more curved the space, the slower the passage of time. On Earth, this effect is pretty much unnoticeable; the difference between the (slower) flow of time at the floor and at the ceiling of a room (gravity is weaker as you move away from Earth) is less than one part in a thousand trillion. However, close to a very massive star, or a compact object such as a white dwarf or a neutron star, the effect is much larger. And close to a black hole, time comes practically to a standstill, as we will soon see. That strong gravity slows down the passage of time is not easy to see. Here is one illustration of this phenomenon, which, like many others, gives only a general idea of what truly goes on. Imagine a box full of atoms, which are emitting radiation at a certain frequency as measured on Earth. Say the strongest emission line is a tone of blue. Since frequency is just a count of how many wave crests pass by a point per second, we can think of the atoms as clocks. Now you place the box close to a neutron star and go far away before you check the emission lines coming out of it. Moving far away guarantees that, unlike the atoms in the box, you are not being strongly influenced by the gravitational field of the star. The blue emission line is gone, and you see instead an orange line. If you place the box closer to the star, the line turns redder. You conclude that the stronger the gravitational field, the more the radiation is shifted toward the red—that is, the smaller its frequency and thus its energy. Close to the neutron star, the photons have to work much harder to escape the stronger gravitational pull, wasting some of their energy as they "climb" up, like children trying to climb up a slide "the wrong way"; the steeper the slide the harder the climb. This effect is known as *gravitational redshift*. Since I argued that you may think of vibrating atoms as clocks, a strong gravitational field decreases the frequency of its emitted radiation and thus the flow of time; a "second" measured by a clock hovering near the star lasts longer than one measured far from it, independently of the kind of clock being used. We can thus imagine a gravitational field so strong that photons lose all their energy trying to climb out of it—light gets effectively trapped by the gravitational field—and time comes to a standstill; this is the exotic reality in the neighborhood of a black hole.

Let me sum up. Einstein equated gravity with the curvature of space around a massive body. The effect is quite negligible for light masses, but becomes important for massive stars and even more so for very compact objects such as neutron stars, where gravity's pull is one hundred thousand times stronger than at the Sun's surface. As small masses are placed in the neighborhood of a larger mass, the distortions of space around it will cause their motions to deviate from what is predicted by the Newtonian theory. Another remarkable consequence of Einstein's new theory of gravity is the slowing down of clocks in strong gravitational fields. This effect is related to the redshifting of emission lines from atoms; photons must work harder to escape the gravitational pull of strong fields, losing energy and thus decreasing their frequency, according to the predictions of general relativity. In short, strong gravity bends space and slows time. We are now ready to explore black holes, the cosmic maelstroms.

Descent into the Maelstrom: Prologue

Dark water is ominous; I know excellent swimmers who refuse to dive into the dark waters of a lake. There is a sensation of loss of control, of not knowing what could be underneath, just waiting to get you. If the waters of Loch Ness were crystal clear, there would be no stories of a monster in them. The mystery is in what you cannot see, but only imagine. And if there is motion in the water, things just get worse. Especially if this motion is a whirlpool, dragging everything that comes too close to it down its deep, dark throat toward . . . who knows what? Medieval and Renaissance chronicles tell of a giant whirlpool off the coast of Norway, referred to as the *gurges mirabilis,* the wondrous and terrifying Maelstrom, the *umbilicus maris,* navel of the sea.[7] The power of this whirlpool was legendary. Some believed the Maelstrom to be the main gate controlling the ebbing and flowing of the tides of all oceans, which were connected by underground tunnels. Athanasius Kircher (1601–1680), the German Jesuit and scholar sometimes called the "last Renaissance man," described in his *Mundus Subterraneus* (1665) how "every whirlpool formed around a central rock: a great cavern opened beneath; down this cavern the water rushed; the whirling was produced as in a basin emptying through a central hole"[8] (see figure 34 in insert). This seventeenth-century description has many of the central elements that will help us visualize what happens near a black hole: whirlpools forming around a central object; a great cavern into which matter rushes; a central hole through which things seem to disappear.

In Norse mythology, as in several others,[9] the Maelstrom was believed to be the passage to the world of the dead, a tunnel connecting the living and the nonliving. Whoever ventured—or was sucked—down its monstrous throat would never come back alive. Well, one person did come back from the depths of the whirlpool, the tortured narrator of Edgar Allan Poe's short story, "A Descent into the Maelstrom," a man whose jet-black hair turned completely white during his horrifying experience. How long his plunge into the whirling abyss lasted we are not told, because the fisherman's watch stopped ticking as soon as he crossed the imaginary line of no return, the edge of the swirling current. But he does tell us how terror changed to awe as he faced the stark beauty of the phenomenon, and realized "how magnificent a thing it was to die in such a manner, and how foolish it was in me to think of so paltry a consideration as my own individual life, in view of so wonderful a manifestation of God's power." After this cathartic revelation, our narrator fell prey to an irresistible desire to know what lies beyond the dark abyss, a feeling I equate to the curiosity that drives scientists and explorers to probe unknown worlds, be they part of our physical reality or an imaginary one: "After a little while I became possessed with the keenest curiosity about the whirl itself. I positively felt a *wish* to explore its depths, even at the sacrifice I was going to make; and my principal grief was that I should never be able to tell my old companions on shore about the mysteries I should see." It is time our narrative starts, a fictional descent into a black hole, wherein some of the physical properties of these strange objects will be illustrated while others will merely be fantasized.

Descent into the Maelstrom: A Fantasy

In my young days, when time seemed to be an endless commodity, I used to build spaceships out of scrap parts of old ones. My intent was not so much to build a functional interstellar cruising machine as to pay homage to the pioneer days of space exploration, when, thousands of years ago, we first attempted to leave our original solar system in search of alternative homes. In retrospect, we should have listened to the scientists when they forecast the catastrophic consequences of climatic change owing to global warming. It was thus a hobby of mine to travel from planet to planet looking for old spaceship repositories, where all sorts of bits and pieces could be found. On one of my travels in search of a rare gyroscope for a 2180 Mars Lander, I found "Mr. Ström's Rocket Parts," an enormous hangar littered with countless spaceship parts. While I was consulting the store's virtual stock-scanning device to search for the gyroscope, Mr. Ström himself came to greet me. I had heard of Mr. Ström before, as had many other space buffs; he was famous throughout the galaxy for claiming to be the one who came closest to a black hole, a story that, to most people, was just that, a story.

Like many before me, I asked Mr. Ström to tell me his story. As he stared at me, trying to read my true intentions, I realized how devastating it must be to have your story discredited by everyone, to be a forsaken hero, whose feats no one believes. It was no wonder his face looked like a fortress abused by the hands of man and time, crisscrossed by many wrinkles that told of countless misadventures and secret pains. His eyes truly made me shudder—two dark pools where only the deepest sadness could swim, twin abysses luring you deeper and deeper into oblivion, like miniature black holes.

"I was commander of a fleet built to explore the complex astrophysical x-ray source known as Cygnus X-1," he started. "Since the 1970s, over three millennia ago, this was suspected to be a binary star system six thousand light-years away from Earth. The two members of the binary system, thought to be a blue giant star about 20–30 solar masses and a black hole about 7–15 solar masses, orbited so close together that the black hole frantically sucked matter from its huge companion, very much as an emptying drain sucks water into a spiraling oblivion. This mad swirling heated the infalling stellar matter to enormous temperatures, producing the x rays observed on Earth by astronomers. Even though observations indicated that the smaller object of the pair had a mass much larger than the maximum mass for neutron stars, it was still not clear whether or not it was a black hole. Since other attempts to identify it as such had failed, the League of Planets decided that the only way to know for sure was to go there.

"The fleet consisted of three vessels, each under the command of a Ström, a great honor to my family, dedicated to space travel and exploration as we had been for many centuries. I led the vessel named *CX1,* my middle brother led *CX2,* and the youngest led *CX3.* Had I known what was about to happen, I would have made sure the three of us were all together in one vessel. . . . I will spare you the details of how the mission was prepared, and how, after many problems with our hyperrelativistic plasma drive, we finally arrived to within one light-month of our destination. [For fictional purposes, let us assume that, somehow, in three thousand years it will be possible to cover vast interstellar distances in reasonable times.] It was a majestic sight the likes of which I had never seen; through our telescopes, we could see an enormous hot blue star, being drained of its substance by an invisible hole in space! All you could see was the matter swirling around this very small region, emitting all sorts of radiation as it plunged into the dark abyss, electromagnetic shrieks of despair if you ask me.

"We were instructed to fly Indian file toward the black hole, keeping a very large distance from each other, my youngest brother first, my middle brother second, and I last. We knew that, from a distance, a black hole behaves like any other massive object, that the differences predicted by general relativity happen only fairly close to it. We also knew that every black hole has an imaginary limiting sphere around it known as the 'event horizon,' which marks the distance from which not even light could escape. This distance is also known as the Schwarzschild radius, since it was Karl Schwarzschild who in 1915, just months after Einstein published his new theory of gravity, first came up with a solution to the problem of the gravitational field around a massive body. Every object, celestial or not, if compacted enough, can turn into a black hole with its own event horizon; for the good old Sun to turn into a black hole, it has to be shrunk into a ball with a radius of 3 kilometers, its Schwarzschild radius, while a human must be shrunk to 10 trillion trillionths of a centimeter [10^{-23} cm]."

"Yeah, yeah," I said, my impatience betraying my excitement, "I also took intro physics in junior high. Please go on with your story!"

"Very well," Mr. Ström whispered, his dark eyes inexorably pulling me in. "My youngest brother's ship, the *CX3*, was to approach the hole, sending us periodic light flashes of a given frequency; we were to follow at a distance, measuring the frequency of the radiation emitted by my brother's ship as well as the time interval between the flashes, and then compare them with the theoretical predictions for gravitational redshift and time delay.* The three vessels plunged to a distance of 10,000 kilometers from the hole [by comparison, Mercury is at an average distance of 58 million kilometers from the Sun]; while *CX1* and *CX2* hovered at that distance, my brother closed in to 100 kilometers from the hole. He was instructed to send us infrared radiation; but we detected only low-frequency [invisible] radio waves. The gravitational redshift formula was indeed correct. Furthermore, the intervals between two pulses increased quite perceptibly; time was flowing slower for my brother, as viewed from our distant ships. He plunged to the dangerously close distance of 10 kilometers from the hole, only 7 from the event horizon; this was the closest distance the [fictionally sturdy] ship could stand, given the enormous tidal forces around the hole, which stretch everything into spaghetti. A French astrophysicist once compared the stretching forces on a person standing up at the event horizon of a 10-solar-mass black hole to what he would feel if he were hanging from the Eiffel

---

* For simplicity, I will take the mass of the black hole in Cygnus X-1 to be 1 solar mass, although we know it to be much larger, somewhere around 10 solar masses. (We are not sure at present, i.e., 2002 C.E.)

Tower with the entire population of Paris suspended from his ankles.[10] Ouch! Anyway, from that close orbit, my brother was to send pulses of visible light, but all we detected were even lower-frequency radio waves; we could not see my brother's ship any longer, and I started to feel very uneasy about this whole thing. Yes, the theory was correct—a ship falling into a black hole will become invisible to people in a more distant ship (us) because of the redshifting of light. That also meant that we would never be able to actually see a star collapsing into a black hole, since it will become invisible before it meets its end. A related effect was the slowing of time. As my youngest brother approached the black hole, the radiation pulses were arriving at increasingly long intervals. Thus, not only could we not see him anymore; we would also have to wait an enormous amount of time to receive any message from him. This confirmed the prediction that, indeed, for a distant observer the collapse of a star would take forever. Of course, for the unlucky traveler who free-falls into the black hole, nothing unusual with the passage of time would happen, as is explained by the equivalence principle; gravity is neutralized in free fall. However, his body would be horribly stretched."

"Yes, I know that," I said, shifting nervously on my chair. "Please do go on!"

"But the theory said nothing about steering a spaceship around the enormous gravitational pull of a rotating black hole, which makes the horrible currents around the fabled Maelstrom look like waves in a child's pool. The orbital instabilities were compounded by the enormous amount of turbulence created by the swirling matter of the blue giant. Still, the radio signals from my brother kept coming in, albeit at very large intervals. His instructions were to return to us after completing two orbits around the hole. More than that, and the spaceship would not be able to take the major tidal stresses and the continuous bombardment of material falling into the rotating abyss. Another message came in, which we decoded as saying, 'MS. Bottle—End.' My middle brother, a great fan of ancient gothic literature, understood it immediately; the message was asking us to read the last paragraph of Edgar Allan Poe's short story 'MS. Found in a Bottle':

> Oh, horror upon horror! . . . We are whirling dizzily, in immense concentric circles, round and round the borders of a gigantic amphitheatre, the summit of whose walls is lost in the darkness and the distance. But little time will be left me to ponder upon my destiny! The circles rapidly grow small—we are plunging madly within the grasp of the whirlpool . . . oh God! And—going down!

My youngest brother was dragged in by the furious rotation of space itself around the black hole. How ironic that words written in 1833 would be so true three thousand years later, around an object too terrifying even for Poe's imagination." Tears of longing and helplessness flooded Mr. Ström's deep dark pools. But they vanished as quickly as they appeared.

"I decided to try and find my youngest brother. Who knows? Maybe the theory was wrong, and it was possible to plunge within the event horizon and still somehow escape. The theory predicted that, once you cross the event horizon, the only possible movement is straight into the center of the hole, the so-called central singularity. No matter what you do to escape, you can't. It is as if inside the event horizon space becomes unidirectional, all roads pointing toward the center, just as time is unidirectional outside the hole, always pointing toward the future. At the singularity, the gravitational pull is infinitely strong. Now, as you know, whenever a physical theory predicts that some quantity is infinite, it is often also predicting its failure to explain what is truly going on. In a sense, the singularity signals the breakdown of general relativity, and is begging for something else to come in. Many physicists believe that this something else is quantum mechanics or, better, a theory that weds quantum mechanical ideas to general relativity. This idea appeared during the twentieth century and is still with us, unproved but on much firmer ground.* Since we know that the behavior of matter at atomic and subatomic scales is very different from that at human scales, we predict that the properties of space and time will also change radically at extremely small scales, much smaller than an elementary particle. This is where a theory of *quantum gravity* comes in, trying to make sense of what happens to space and time at extremely small distances. But for us, human-size black hole explorers, these quantum effects are of no use; as we plunge in, the gravitational monorail will crash us into whatever lies at the core of the hole, and we will not be able to deviate from it or send signals outside to tell our story. Unless, of course, the hole is rotating. And this one was."

My ears perked up, for I sensed that something wonderful was about to be revealed to me. I knew that the crushing central singularity existed only for static black holes, those that do not rotate. Since rotation is ubiquitous in the universe, most black holes would form from rotating stars and would also

---

* I believe and hope that in three thousand years we will know how to wed the two theories! However, for dramatic reasons—and to avoid speculations that will probably be incorrect—I am assuming the question will still be open that far in the future.

rotate. Now, I also knew that rotation has the effect of prying the central singularity open, distorting it into a ring. During the twentieth century, physicists conjectured that—mathematically, at least—it was possible to construct solutions of Einstein's equations where a rotating black hole terminates not in a central singularity but in a throat connecting to a white hole, its exact opposite! What the black hole takes in, the white hole spits out. If they exist, white holes would be sources of matter and radiation at another point of the universe or even another universe, although here fanciful speculations blur proper judgment more and more; try as we want, not every compellingly beautiful mathematical result has a place in nature. The existence of white holes has remained unclear for all these years, as no conclusive proof or observation has been offered so far. The throat connecting a black hole to a white hole (maybe a "white source" would be a more accurate, if less appealing, name) is called an Einstein-Rosen bridge or, when connecting two different points in our universe, a wormhole (see figure 18). In fact, back in the 1980s, the astronomer Carl Sagan wrote a novel, *Contact*, in which the heroine communicated with an advanced alien intelligence through wormhole connections. Inspired by this and other stories, we have spent three thousand years looking for wormholes and trying to figure out how to keep their mouths open; for the theory also says that keeping a wormhole mouth open requires very exotic kinds of matter, which exert a sort of "antigravity" effect on ordinary matter. Could Mr. Ström's experience shed some light on this theoretical impasse?

"I decided to risk my life and go into the maelstrom after my brother," Mr. Ström continued. "My middle brother waited in a safe distant orbit around the black hole. He was to report to base within three days, irrespective of what happened to me. As I plunged into the hole, the whirling of space dragged me in the way water drags a ship into a whirlpool.* The combination of enormous gravitational pull and furious bombardment of radiation and particles took a toll on my ship; but its fuselage miraculously—what else could it be but a miracle?—survived, as I did, thanks to the once controversial anticrunch shield. I couldn't see a thing, but my brain was producing all sorts of images and sounds, as if all my neurons decided to fire in unison. I saw my past life and what I believed would be my future life. But this image bifurcated into countless others, each spun by a choice I made along the way, possible destinies wrapped into an instantaneous glimpse. Strands of time were rolled up in a ball, which I could hold in the palm of my hand, while space convulsed into infinitely many

---

* Caution: What follows is *completely* fictional. It is a playful game inspired by speculations in general relativity.

shapes and forms coexisting in one single point. My mind envisioned all that was imaginable, possible or impossible. It became clear to me that what we call impossible depends on where we place the boundaries of reality. And at that moment, reality had no boundaries. I saw my dead relatives and those who were not yet born; I saw myself, as an adult, talking to my mother, even though she died when I was a child. As she sadly apologized for her absence, for not being able to see me grow, or to give me her love, she became the child and I the adult; and for the first time I understood she had no choice. We were standing in the living room of our old home but also flying over deserts and swimming underwater. I saw myself being born, and my mother saw me dying. Our images merged into a knot, which became a tear in my mother's eye. As I raced to embrace her, desperate for her touch after so many years, the spaceship was flooded by the most intense light I had ever experienced. It was too late. My stretched-out hands passed through her body as she vanished into bright nothingness. I was alone again, blinded by whiteness, my innermost hopes defeated. The black hole had fed on my dreams and mocked my pain.

"I felt an enormous push, as if the spaceship were being coughed up by a giant. The last thing I remember was hitting my head against the wall. I must have remained unconscious for quite a while, but could not be certain, because all my clocks were damaged. When I looked in the mirror, I could hardly believe what I saw; my hair had turned completely white and my face was covered with wrinkles I didn't have moments (moments?) ago. I checked my location in the computer and realized that, somehow, I reemerged two thousand light-years away from Cygnus X-1! The only possible explanation was that I traveled through a wormhole that somehow was kept open inside the black hole and was tossed out by a white hole at a faraway point in space. I had glimpsed eternity and contacted my dead loved ones, I had held infinity in my hands, but was unable to find my youngest brother. And yet, to this day I am certain that he also survived, that he reemerged somewhere in this universe, taking a different path through the wormhole. I feel that we remain connected by threads leading back to the maelstrom, a place where all the choices we made coexist with those we didn't, where all our I's are possible."

I was deeply moved by Mr. Ström's narrative. Its truth was irrelevant to me. Somehow, the image of all my life coexisting in one point in space and time became more important than its plausibility. The black hole pilgrimage, for it was a pilgrimage, was an expression of Mr. Ström's innermost desires, a search for the totality of his self; science had given him the possibility of contemplating immortality. Even beyond that, I realized that science allowed him to imagine a place where time could flow in any sense, or many senses at once, or not at all.

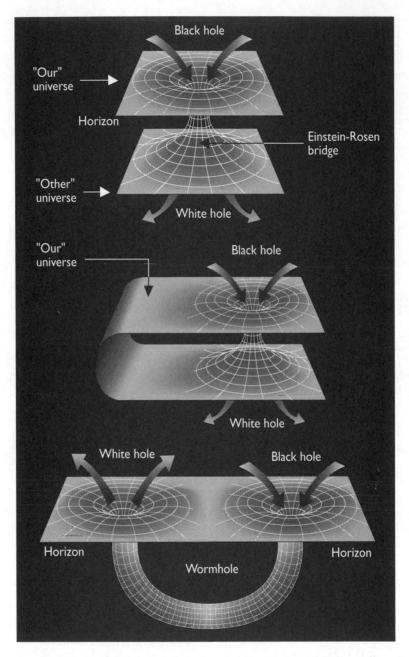

FIGURE 18: Top: *A two-dimensional illustration of an Einstein-Rosen bridge, connecting first two points in hypothetical different universes, or two points in our own.* Bottom: *A wormhole connecting two points in our universe. In both cases, one goes in through a black hole and out through a white hole.*

Who knows? Maybe this place really exists, and Mr. Ström's narrative is factual. Or, if it doesn't, one day we may create it. For it is precisely at the cutting edge of knowledge that imagination is most crucial; what was once only a dream may very well become a reality tomorrow.

Descent into the Maelstrom: Postscript

We have traveled far into the realm of the scientific unknown, where what we can take for certain blurs with mathematical speculation. Our present knowledge of white dwarfs, neutron stars, and black holes may not be complete, but it certainly is very extensive. This knowledge has transformed the way we see the cosmos and its many worlds, confirming once more the amazing and limitless creativity of nature. Kant would have been ecstatic to see his image of the "Phoenix of Nature," of repeating cosmic cycles of creation and destruction, animated so spectacularly in the birth and death of stars. But he would perhaps have been incredulous to see that ideas dating back to his contemporaries John Michell (1724–1793) and Laplace, of stars so massive as to be capable of swallowing their own light, were also vindicated. And he would have been even more incredulous to learn that those objects could turn Newtonian notions of absolute space and time, so precious to his philosophy, inside out.

The event horizon separates our universe, or what we can observe, from the central singularity of a black hole, protecting us from whatever lies inside. This protection has been dubbed cosmic censorship, as if nature itself were protecting our limited reasoning abilities from the bizarre reality that lies within. The possibility of actually having a spaceship travel through the throatlike singularity of a rotating black hole, as did our hero Mr. Ström, or through a wormhole in space, is at best very remote. The possibility that this traveler would actually reemerge unscathed somewhere in the universe is even more implausible. We don't know whether these cosmic shortcuts indeed exist, and, if they do, how they could be kept open, and whether they could be a passage to somewhere else in space and in time. But ignorance and implausibility should never stop us from speculating, so long as the speculation is grounded on solid scientific reasoning. This is what many astrophysicists working on black holes do, and it is thanks to their intellectual boldness and creativity that we have learned so much about these fascinating objects. We don't know whether science can provide secret passages to other realities, like the closet in *The Lion, the Witch, and the Wardrobe*, by C. S. Lewis, or the tree in *Alice in Wonderland*, by Lewis Carroll. But if a scenario can be reasoned through the laws of physics, the effort is worthwhile. As the narrator of our fictional tale said, "it is precisely at the cutting edge of knowledge that imagination is most crucial; what was once only a dream may very well become a reality tomorrow."

We now believe that most galaxies nest a giant black hole in their center, monsters of millions or even billions of solar masses. A black hole with three million solar masses, like the one we believe exists in the core of the Milky Way, has an event horizon of about 8 million kilometers, twenty times smaller than the distance between the Sun and Earth, tiny in comparison with the galaxy. Thus, we don't have to worry about being sucked down this black hole anymore than of being dragged down by a whirlpool in the Caribbean while swimming in Rio. With hundreds of billions of galaxies out there, each with hundreds of billions of stars, we do have an enormous sample of black holes to study, from small-solar-mass ones to true cosmic giants. And study them we will, as several missions are being planned with orbiting x-ray telescopes to image known black holes in unprecedented detail, upgrading the current Chandra satellite from NASA. Within two decades, we will know much more about these objects and their properties. Will we know what lies inside the event horizon, or whether wormholes do exist? That is hard to guess. My own feeling is that at most we will know better how plausible our present speculations concerning the interior of black holes and cosmic shortcuts are and that we will create many more guesses. But then again, I could be wrong . . .

We will now take the last logical step of our voyage, from stars and black holes to the universe as a whole. In discussing the possible end of Earth and of our Sun, we encountered a new physics, where space and time gain a plasticity that reaches a mind-boggling climax inside black holes. But planets, people, and stars exist *within* the universe and take part in its history. And what history is this? We are confident that the big bang model of modern cosmology provides at least a broad-brush description of the universe, from its beginning to its "end." In the last part of this book, I will present our current ideas about the fate that awaits the universe. After all, this fate is, in a very deep sense, our own. Will it end, as the American poet Robert Frost once wrote, in fire or in ice? For the first time in history, we believe we actually know the answer. Or almost. This cosmological discussion of possible ends will take us back full circle to our opening reflections on time and immortality. Under the ever vigilant eye of science, of course.

PART IV

To the End
of Time

CHAPTER 7

Fire and Ice

Some say the world will end in fire,
Some say in ice.

— ROBERT FROST

Right behind the Physics and Astronomy Building at Dartmouth, surrounded by towering pine trees, there is a life-size bronze statue of the American poet Robert Frost. He sits quietly on top of some rocks (granite, of course), notepad on his lap and eyes lost in thought. Few poets, if any, have captured the essence of life in rural northern New England, with its hardships and stark beauty, as did Frost. The bronze celebrates Frost's genius and brief passage through Dartmouth, where he was a student for a short period. When my daughter, Tali, was about two, I used to take her for walks around these woods. She was completely enchanted by this shining, motionless man, who looked so real and yet was not. Every time we went back, Tali would immediately dart toward the statue, sit on his lap, and kiss his cheeks. "Why did you kiss the statue?" I would ask in that idiotic adult tone, completely devoid of magic. "Because he looks so sad, Daddy. Is it because he can't move?" To me, Tali's actions captured the essence of our perception of time; time is motion, is how we describe change. A motionless world, a world without time is a sad world, fated to never reinvent itself. Time is the absence of perfection. This, to me, is what the notion of paradise, with its absence of time, implies; perfection is changeless and thus timeless. Eternity is not just an infinitely long time but the absence of time, the absence of change, the absence of renewal, "never-ending present," as Saint

Augustine wrote. It is decreed in the Old and New Testaments that time will come to an end when good defeats evil. While they both coexist, time flows forward, inexorably. We are here, now, dealing with this battle, trying to make sense of our mistakes and choices. What would life be if we made no mistakes and thus needed no forgiveness? Would we still be human? I, for one, am not ready to give up time.

Frost, like most of us, though more eloquently, wondered about the end of time. His well-known poem "Fire and Ice" appeared in *Harper's Magazine* in 1920, two years before the Russian meteorologist turned cosmologist Alexander Alexandrovich Friedmann obtained the first cosmological solutions showing that the universe as a whole could indeed change in time, growing or contracting, and actually ending in fire or in ice. Frost's inspiration probably came from combining two more Earth-bound ends, the apocalyptic scenario of Christianity—falling stars and all-consuming fire and doom—with the desolation of a New England frigid winter and the possibility of a new ice age. However, Frost's description could hardly have been more appropriate for the cosmology of his days, eerily timely and prophetic. In 1917, two years after publishing his general theory of relativity, Einstein showed that his equations applied not only to stars but to the universe as a whole; just as we need the matter distribution of a star to obtain the geometry of space inside and around it, if we know the matter distribution in the universe we can obtain its geometry. That must have been a very powerful feeling, when Einstein first realized that he could actually determine the geometry of the universe through physics and mathematics.

During the past eighty years, cosmology has gone through some very deep transformations, changes, and revisions. Most important, it has become an experimental science, as opposed to a merely mathematical exercise inspired by general physical concepts. The desktop universes of the early twentieth century, inspiring as they were, are gone for good, thanks to an enormous joint effort of cosmologists and astronomers. But even though their work has solved many conundrums, it has also given rise to many more. We are now in a paradoxical age where our successful observations have introduced tremendous uncertainties about the cosmos. We don't know what most of the universe is made of. We don't even know whether its dominant energy contribution comes from material particles or from some sort of diffuse "vacuum energy" being pulled from the quantum world. We also don't know the universe's fate and may not even be able to predict it. And yet, we have never known as much about the universe as we do now!

This strange state of affairs reminds me of the "crisis" in physics during the late nineteenth century, when widespread theoretical confidence—bordering on arrogance—was being undermined by a series of experimental discoveries that just didn't fit the classical worldview so carefully constructed since the days of Newton. A new revolution was about to happen, the quantum revolution that profoundly changed our view of the atomic and subatomic worlds, redefined our destructive capabilities (through the use of nuclear energy), and, to a large extent, triggered the information revolution of the late twentieth century, with its semiconductors, lasers, and digital devices.[1] It is quite possible that we are about to witness a similar revolution in our understanding of the cosmos. Some scientists even say that we are already in the midst of it. Our confusion couldn't be more stimulating! In this chapter you will see why.

The Cosmos, circa 1945

Twentieth-century cosmology did not advance at a steady pace. Periods of focused activity were followed by times of relative calm and isolated interest. It is fair to say that, until the 1960s, most physicists viewed cosmology with great suspicion. It is also fair to say that, unfortunately, this attitude remains intact in some circles, although these are rapidly dwindling. Two reasons mostly account for this suspicion. First, cosmology asks the big questions—"How did the world come to be?" or "Will the universe end?"—which have traditionally been the province of religious and metaphysical inquiry. This unavoidable link has long distinguished cosmological research from more "down to earth" questions related to the behavior of localized systems, whether subatomic particles, crystals, or black holes. Second, until the mid-1960s there was an appalling lack of data backing up or contradicting the many desktop universes proposed earlier on. How could cosmologists claim to be doing science without having their hypotheses tested? After all, without observational evidence, the only guidance we have in building models and hypotheses is mathematical elegance and physical intuition, which are crucial but not definitive. What may be mathematically compelling may not necessarily correspond to physical reality. In all honesty, cosmologist that I am, I must admit that the skepticism on the part of most of the academic community was partly justified. But not any longer.

Einstein's pioneering cosmological research was based on a general principle still crucial to cosmology today, the *cosmological principle*: on the average, the universe looks the same everywhere. In other words, all points in the universe are essentially equivalent. Two images often used are the perfectly smooth surfaces of a ball and of a very large square table, both examples of two-dimensional geometries where all points are equally important. (For the table, we must be sufficiently far from the edges, since they are special. Only if the table were infinitely long in all directions would its flat geometry truly satisfy the cosmological principle.) A quick glance at a moonless night sky shows that this principle should not be applied to the local cosmic neighborhood we can see with our naked eye; stars are distributed unevenly and do not hint at a cosmic homogeneity. To apply the cosmological principle, we must move toward much larger distances, not of a few tens of light-years, but of hundreds of millions of light-years. In fact, the precise distance scale at which we can actually say the universe is on the average the same has steadily grown, as increasingly large complex structures—huge clusters of galaxies and huge empty regions called voids—were discovered during the past two decades. But we now think we have arrived at a consensus that, at these enormous distances, the universe can indeed be interpreted as being homogeneous.

In practice, we can illustrate the cosmological principle by means of the vegetable soup approach; if we put a bunch of vegetables in a cooking pot filled with water, spices, beets, carrots, potatoes, and the like, the final soup will be very inhomogeneous, that is, lumpy. So, to make it homogeneous, we put the cooked veggies in a blender until they are transformed into a smooth puree. The final creamy soup will have the same amount of matter (mass) as the lumpy soup, but with the crucial difference that it will satisfy the inside-the-pot cosmological principle, that all points within the soup are essentially equivalent. Einstein put the universe in a blender, so to speak, and assumed that the geometry of the cosmos would be determined by the amount of matter within it, distributed homogeneously as in the smooth vegetable soup. This simplification allowed him to find a solution to his equations describing a static universe. Why static? In 1917, Einstein had no reason to suppose that the universe was a dynamic entity that could actually evolve in time: the expansion of the universe was not conclusively discovered until 1929. Einstein was, of course, very proud of his solution, even though the requirement of having a static cosmos forced him to include an extra term into his equations, which he called "negative pressure." (Reader, take note of this, because negative pressure will come back full force in late-twentieth-century cosmology.) We now call this term a *cosmological constant,* a mathematically allowed contribution to the equations of general relativity that acts as a kind of antigravity, pushing the geometry apart. Remember that the matter (and thus energy) distribution determines the geometry. Einstein needed the cosmological constant in order to stabilize his universe; a static universe filled with a homogeneous distribution of matter would spontaneously collapse upon itself because of its own gravity. The existence and nature of this inelegant intruder, which efficiently performed the balancing act needed in Einstein's universe, remains one of the greatest mysteries of modern physics.

Friedmann freed Einstein's universe from the chains of immobility by allowing matter to change in time. And since general relativity states that matter determines the geometry, if matter changes in time so does the geometry. Before discussing Friedmann's solutions, I will introduce a concept crucial to cosmology, the *energy density*. The quick, dirty, and, for us, sufficient definition is that energy density is the total energy divided by the volume that contains it; if you have a certain amount of energy in a cubic box and you decrease the sides of the box by half, the energy density grows by a factor of eight;* in a smaller box, the energy is squeezed together and hence its density is higher. If you double the sides of the box, the energy density decreases by the same factor of eight. Thus, in an expanding box the energy density decreases, while in a contracting box it increases. The key question, which we will revisit as we trace the development of modern cosmology, is what contributes to the energy density of the universe.

There are three possible contributions to the energy density of the universe, which may appear individually or together. There may be a *radiation* contribution, consisting of photons, the particles of electromagnetic radiation we encountered before, and any other particles with no mass or very low mass, such as neutrinos. *Matter* also contributes, since it carries energy in its mass—through the $E = mc^2$ relation—and in its motion; here we must think of matter not just in the form of galaxies and stars but also in a more elementary form, that is, broken down to more fundamental constituents. What these constituents are—protons, neutrons, or something more exotic—will depend on what kind of matter fills the universe. Finally, the *cosmological constant,* or some other diffuse form of energy that mimics its effects, may also contribute to the geometry of the universe.

Several desktop universes were found during the 1920s and 1930s, by solving Einstein's equations with varying kinds of contributions to the energy density. Apart from Friedmann, other key players in this game were Arthur Eddington, whom we met in chapter 6, Georges Lemaître, a priest cum cosmologist from Belgium, and the Dutch Willem de Sitter. The key broad-brush

---

* Here is why: the density (D) of energy (E) in a box of volume V is $D = E/V$. For a cubic box of size L, $V = L^3$. Thus, if you halve the side, $D = E/(L/2)^3$. Since $2^3 = 8$, D gains a factor of 8.

feature of these solutions (some subtleties will be dealt with later) is that they exhibited two possible behaviors for the geometry of the universe: eternal expansion or a cyclic sequence of expanding and contracting eras, the so-called Phoenix universe. (Kant would have enjoyed this tremendously!) In the absence of a cosmological constant, the key parameter determining the fate of the universe—what my colleagues Lawrence Krauss from Case Western Reserve University and Michael Turner from the University of Chicago aptly and eschatologically called the relation between geometry and destiny—is the energy density.

The cosmological equations determine a critical value for the total energy density—that is, the sum of radiation, matter, and other contributions—which, in the absence of a cosmological constant, seals the fate of the universe; they put all contributions to the energy density of the universe into Einstein's cosmic blender, so as to make it homogeneous throughout. The all-important critical energy density is about one atom of hydrogen in a cube of roughly half a meter a side. If the total energy density is above this value, the gravity of the universe will eventually pull it back and cause it to collapse into a "big crunch"; if the energy density is below this value, the universe will keep expanding forever. It will also expand forever if the energy density is *exactly* equal to the critical value. A very useful analogy is a rocket being launched from Earth. The engines thrust the rocket up, and Earth's gravity pulls it back down. Clearly, if the rocket reaches a high enough velocity, it will keep going up and will eventually escape Earth's gravitational pull. (It will still be pulled by Earth's gravity, but to very small effect, since gravity decays with the square of the distance.) Otherwise, it will eventually reverse its motion and come crashing down on the ground; the closer the rocket is to this critical velocity (called escape velocity), the higher it will travel before it starts falling back. You can easily verify this by throwing a rock upward with different speeds; the greater the speed, the higher the rock travels before coming down. In fact, if you can throw it at 11 kilometers per second, about 25,000 miles per hour, the rock will escape into outer space!

Now, imagine that the rocket can travel only at a fixed velocity. This rocket will be able to escape planets only if its velocity is equal to or greater than the planet's escape velocity. For example, a rocket that can narrowly escape from Earth will be equally successful escaping from Mars, but will fail miserably trying to escape from Jupiter. The lesson is simple; the larger the planet's gravitational pull, the harder it is to escape from its surface. This is where the analogy with the universe gets more interesting. Why does the universe expand in the first place? After all, the rockets have engines to push them up. The reason is that in very early times the universe was so small that matter and radiation were squeezed to absurdly huge pressures and temperatures. Think of squeezing down some very powerful springs and letting go—it was much like that with the universe.

But this is not the whole story. Apart from the energy density, the equations that control the expansion of the universe have a second term, determined by the geometry of space. There are three possibilities: a *flat* geometry, a three-dimensional analogue of the flat table we discussed above; a *closed* geometry, a three-dimensional analogue of the surface of the ball (it is pretty much impossible for humans to visualize this, hence the two-dimensional examples); and finally, an *open* geometry, which in two dimensions we can visualize as a saddle, curving in two opposite directions. For a flat geometry, the spatial curvature term is absent from the cosmological equations and the universe expands at its escape velocity. For curved spaces, however, it is there and it makes a very big difference. If the geometry is closed, the universe will never be able to expand forever; eventually, the curvature term will dominate and will force the universe to collapse on itself. Imagine attaching a spring to the rocket; as it tries to escape the gravitational pull, the spring eventually gets stretched to its maximum length and starts pulling the rocket back down. Likewise, a closed universe without a cosmological constant will reach a maximum radius and collapse upon itself. A closed geometry seals a destiny of collapse.

If the geometry is open, the universe will be underdense and will expand forever; the open geometry gives an extra "kick" to the expansion, the exact reverse from the overdense closed-geometry case. Back to our rocket analogy: the spring attached to the rocket was manufactured to keep stretching indefinitely as it expands, getting weaker as it gets longer, boosting the rocket's escape. An open geometry seals a destiny of continuous expansion (see figure 19).

To summarize, in the absence of a cosmological constant, there is a one-to-one correspondence between the energy density of the universe, its geometry, and its fate:

Undercritical density => open geometry => expansion
Critical density => flat geometry => expansion (borderline case)
Overcritical density => closed geometry => eventual collapse

But these, of course, were all desktop universes, since physicists during the 1920s had no conclusive observation as to the dynamical behavior of the cosmos. This situation changed dramatically in 1929, when the American astronomer Edwin Hubble discovered that the universe is expanding in accordance with a

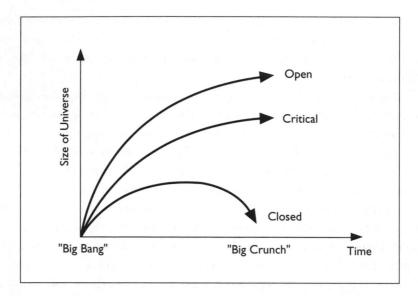

FIGURE 19: *The three possible fates of the universe, according to Friedmann's cosmology. An over-dense universe has closed geometry and recollapses; a critical universe has flat geometry and expands forever; an underdense universe has open geometry and also expands forever. For Friedmann's cosmology, geometry determines destiny.*

very simple rule; the farther away the object—for example, a galaxy—the faster it recedes from us. In order to come to this amazing conclusion, Hubble needed to measure two crucial numbers: the distance and the velocity of the galaxy relative to us. For the velocity, he used the Doppler effect, the fact that the pitch of a wave rises if its source is moving toward us (or we toward the wave's source) and falls if its source is moving away from us (or we away from the source). We are familiar with the sound waves version of the Doppler effect from our urban and highway experiences with sirens and horns. In astronomy, we measure the changes in the emission spectra of galaxies as compared with those of similar nearby objects; if the galaxy is moving away from us, the radiation it emits will be shifted toward lower frequencies. We say that light from a receding object is redshifted, because tones of red light have the lowest frequencies of the visible spectrum. If, on the other hand, the galaxy is approaching us, its spectrum will be shifted to higher frequencies, or blueshifted. This is the case, for example, of our neighboring galaxy Andromeda, located about 2.5 million light-years away. But how could Andromeda be approaching us, if in an expanding universe all objects at large distances should be getting away from one another? You may think of the cosmological expansion as a river flow, carrying with it all sorts of floating objects. However, here and there two objects may get tangled on a local eddy, remaining temporarily detached from the overall water flow. In the case of pairs or clusters (self-gravitating groups) of galaxies, their local gravitational attraction, if they are sufficiently close to each other, overwhelms the overall cosmic expansion, causing their emission spectra to be blueshifted. Keep in mind that the expansion of the universe is not like an explosion, where everything flies away from a common center, but more like a stretching of the geometry of space. Think of a rubber sheet studded with small rigid disks; if the sheet is stretched, the distances between the disks will automatically increase, as the sheet's stretching carries them away, but the disks will keep their original shape. The same holds for galaxies and other objects drifting in the cosmic flow.

So much for velocities. What about distances? Hubble knew he needed what astronomers call a standard candle, an astronomical object with a well-defined luminosity, which can be identified easily in faraway galaxies. Since, by definition, the "candle" emits the same amount of light wherever it is and its brightness drops as the square of the distance, if we measure the candle's luminosity in the faraway galaxy, we can obtain its distance and hence the galaxy's distance as well. In a terrestrial analogy, we can imagine placing several identical candles at varying distances from our home; by measuring the drop in luminosity of each candle, we can obtain its distance from us. Hubble identified stars called Cepheids, which have a well-known variability in their brightness, in several nearby galaxies. In fact, it was by means of Cepheids that he solved a problem that had been plaguing astronomers for centuries: Is the Milky Way the only galaxy in the universe, or are there other "island universes" out there, as Kant had envisioned? Incredible as it may sound, it was only in 1924 that Hubble showed that the universe contains countless (well, hundreds of billions) of galaxies of many different shapes and sizes. But to extract his velocity-distance relation Hubble had to look deep into the cosmos, and even the 100-inch telescope at Mount Wilson Observatory in California was not powerful enough to find the requisite Cepheids. Hubble then took the next logical step and looked for the brightest stars in each of the distant galaxies, which, of course, were easier to see. Assuming that these powerful icons had approximately the same brightness, Hubble could use them as standard candles and thus extract the distances to galaxies far enough away. The choice of candles varies, but this technique is still extensively used today.

Hubble's discovery of the cosmic expansion changed the course of cosmology. Here was data that conclusively proved that the universe is indeed expanding, as had been conjectured in some of the desktop models of the 1920s. Einstein tossed away his static universe and the cosmological constant in disgust; the universe was a dynamical entity, its geometry being stretched in all directions. But the cosmological constant would not go away so easily. An obvious consequence of Hubble's discovery is that, if the universe is expanding, it was smaller in the past. Using his velocity-distance relation, Hubble estimated that the universe reached a point of enormous density two billion years ago. In other words, Hubble estimated the age of the universe, the time since the initial "bang" that pushed the geometry, matter, and radiation outward. There was a problem, though. It was known then that Earth was older than two billion years. How could Earth be older than the universe? Solutions were quickly proposed, some of them by Eddington and Lemaître, invoking the cosmological constant; for the right choices of the cosmological constant, it is possible to make the universe go through a "coasting phase," in which the distances between objects hardly change while time keeps passing. These coasting solutions allow for a much older universe with distances similar to what we measure today, "solving" the embarrassing age problem. In 1952, Walter Baade—who, as we saw in chapter 6, jointly proposed with Zwicky the existence of supernovae—refined Hubble's distance measurements and showed that the universe was comfortably older than Earth. The cosmological constant could be dropped again. With the outbreak of the Second World War, most cosmological research had come to a halt, as physicists focused on more immediate questions, such as survival, defense, and, unfortunately, attack.

The Cosmos, circa 1980

Consistent as they were with the cosmic expansion found by Hubble, the desktop universes of the 1920s and 1930s focused on what we could call the geometry of the cosmos and not on the physical properties of the matter that filled it. Lemaître, who had an uncanny physical intuition, was the first to realize that the two, geometrical and material history, cannot be separated; one is inextricably related to the other. As a result, it was crucial to know what the stuff was that was squeezed to very high temperatures and pressures early on in the cosmic history. Could the properties of this primordial matter be understood from well-established physical principles? In other words, could cosmologists reconstruct the cosmic history by combining general relativity with atomic and nuclear physics? Questions of this sort launched the second phase of modern cosmology, where the focus changed from devising mathematical models that described the many possible cosmic geometries to physical models that incorporated the all-important influence of the material constituents of the cosmos in its history.

During the early 1930s, Lemaître proposed his prescient "primeval atom" model, where he envisioned a universe evolving from the radioactive decay of a giant atom: in the beginning, there was *the* atom. It was highly unstable, and, as soon as it came into existence (Lemaître did not make clear how this happened), it fissioned into many pieces, and these fissioned into many more. From these fragments, electrons, helium nuclei, protons, and other particles rushed out. By conjecturing that the fragmentation resulted in an increase in volume, Lemaître related the decay of the primeval atom to an expanding universe. He was well aware of the limitations of his ideas, which he considered more a vision than a theory. Nevertheless, he went as far as predicting the existence of "fossil rays," some form of radiation left over from the initial stages of disintegration and expansion. Little did he know that fossil rays were indeed going to be discovered during the mid-1960s.

Eddington, Lemaître's Ph.D. adviser, was not happy. (Between Chandrasekhar, Lemaître, and others, he sure had to deal with quite a few creative youngsters!) He had an aversion to the idea of an abrupt beginning of the universe. "The most satisfactory theory would be one which made the beginning *not too unaesthetically abrupt*," he wrote in *The Expanding Universe*, published in 1933.[2] He went on to suggest a universe where the initial state was given by Einstein's static solution, the one held steady by the cosmological constant. Eddington had shown that this solution was unstable against small fluctuations, like a ball balancing on a narrow fence, which would fall if touched lightly. Thus, he proposed that the universe started as an Einsteinian universe and stayed such for an indeterminate amount of time: "The primordial state of things which I picture is an even distribution of protons and electrons, extremely diffuse and filling all (spherical) space, remaining nearly balanced for an exceedingly long time until its inherent instability prevails."[3] Thus, in Eddington's universe, time starts to tick only when mounting instabilities take over Einstein's universe. These instabilities gradually gather force and eventually propel the universe into its expanding phase, consistent with Hubble's observations. It is somewhat curious that the notion of a universe with a timeless past that "spontaneously" transitions into an expanding era made a comeback in the 1980s, in the hands of the physicists James Hartle and Stephen Hawking, Alex Vilenkin, and Andrei Linde, as a consequence of applying quantum mechanics to cosmology. But we will save this development for later.

A young Russian physicist named George Gamow, who worked with Friedmann until his untimely death in 1925, was listening very attentively to the debates on the primordial history of the universe. Trained as a quantum physicist, Gamow became a master of applying atomic and nuclear physics to cosmology. He was also a master of irreverence, whose puns shook the rigidity of quite a few scientific circles. Apparently, he particularly enjoyed arriving in Cambridge with his very loud motorcycle, which he made sure roared louder the closer he was to "senior" faculty. During the 1930s, he devoted himself to stellar astrophysics, pioneering many of the ideas that eventually led Hans Bethe, William Fowler, Fred Hoyle, and others to obtain the main fusion reactions that ensure a star's balance against gravity, as well as its explosive endings.

Gamow recognized that the primordial universe shared many properties with stars: high temperatures and pressures, matter dissociated into its most basic constituents, and a continuous struggle against gravity's pull. But he was also aware of a key difference: unlike stars, which remain fairly static during their hydrogen-burning stage, the universe expands in time. And as it does so, both the temperature and the density of its material contents decrease. Richard Tolman, the Caltech physicist who helped Oppenheimer with his neutron star and gravitational collapse models, had studied the thermal behavior of the expanding universe in detail during the 1930s. Using Tolman's equations, inspired by Lemaître's primeval atom, and applying his knowledge of nuclear physics, Gamow proposed a cosmological model in which the universe started as a very hot and dense soup of primordial matter, mostly consisting of electrons, protons, neutrons, neutrinos, and, of course, a lot of photons (radiation). Like Lemaître and his primeval atom, Gamow also declined to speculate on where the soup came from.

Gamow reasoned that initially the heat was so intense that every time neutrons approached protons to fuse into a nucleus of deuterium or helium, highly energetic photons would break their nuclear bonds. And if the strong nuclear force had no chance against these furious photons, the much weaker electric attraction between electrons and protons was completely ineffective. But as the universe expanded and cooled below a few thousand billions of degrees, the radiation became less energetic, until it was no longer strong enough to prevent nuclear bonding. This is the beginning of the "nucleosynthesis era," the period between one second and three minutes after the "bang," when light nuclei were hierarchically fused—from hydrogen (protons) to deuterium, to tritium, to helium 3, to helium 4, and so on. Originally, Gamow proposed that all chemical elements were cooked up in the furnace of the primordial universe. At the other extreme, Hoyle and others argued that this was not the case, maintaining instead that nucleosynthesis occurred exclusively during stellar evolution—no primordial nuclear alchemy! Both were right and wrong; Gamow was partly right since light nuclei (up to lithium 7) were fused primordially, and Hoyle was partly right since all heavier elements are synthesized in stars. They were both wrong in hoping that a single physical mechanism could explain all the chemistry of the universe.

Hoyle had a very different cosmology in mind, the so-called *steady-state* model, which posited that the universe never changed in time, being thus eternal and uncreated. In order to reconcile an unchanging universe with the observed cosmological expansion, Hoyle, Hermann Bondi, and Thomas Gold, all from Cambridge University, proposed in 1948 that matter could be created out of nothing *(creatio ex nihilo)* in such a way as to balance the decrease in density caused by the expansion. Hence the name steady-state. Like Eddington, the Cambridge trio had a serious aversion to the notion of a universe that mysteriously and abruptly appeared at some point back in time; it was too close to the Judeo Christian creation event, too close for their intellectual comfort. The same negative reaction was widespread in the Soviet Union, if not among all the physicists, at least among the political authorities.

With his collaborators Ralph Alpher and Robert Herman, Gamow improved his calculations to the point that they made two key predictions. One was that the universe should contain about 75 percent hydrogen and 24 percent helium (according to modern measurements) and a few other light isotopes cooked during the first three minutes. That is, apart from some helium, only about 1 percent of the matter in the universe is actually cooked in stars. The other prediction was that the universe should be bathed in radiation, which should have a current temperature of 5 degrees Kelvin, about -450 Fahrenheit. This radiation, suggestively similar to what Lemaître called fossil rays, appeared when the universe cooled enough, to roughly 6000 degrees Kelvin, enabling electrons and protons to bind and form hydrogen atoms some 300,000 years after the bang. After this time, photons were not energetic enough to interfere with the electrical attraction between electrons and protons, and were free to roam across the universe. In order for faraway objects to be visible to us (optically or in any wavelength), the radiation (photons) they emit must be able to reach us; that is, it must travel fairly unimpeded. Thus, we can think of this all-important change in the properties of the universe, called *decoupling,* as a transition from a primordial era of opacity—when photons were so tightly bound to electrons and protons that they could not propagate far before hitting them—to an era of transparency, where we are until today, some fourteen billion years later. In other words, the decoupling transition marks an absolute

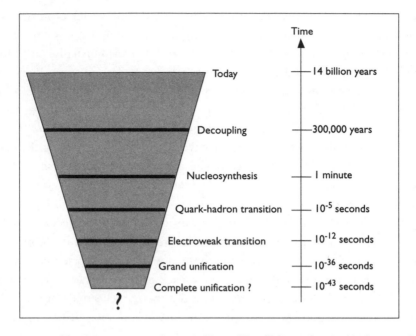

FIGURE 20: *Most important events in cosmic history. We will discuss them in this chapter and the next. (Times are approximate.)*

observational boundary beyond which we cannot probe directly through telescopes of any sort. Any information about the universe before decoupling must come indirectly, through clues left over from these very early times. One example of such a clue is precisely the light chemical elements, cooked when the universe was between a second and three minutes old. That their abundance was predicted by the big bang model and later confirmed by observations gives us confidence that we do indeed understand the physics of the universe at this very early age, a truly remarkable achievement. But we want to go back even further, as close to the initial bang as possible. Perhaps other clues are out there, waiting to be discovered, maybe even encoded somehow in the radiation left over from decoupling. This is what cosmologists have been trying to do since roughly the 1980s, as we will soon see.

The fact that Gamow's model made two testable predictions that could in principle be observed should not be underestimated. This is precisely what distinguishes a physical theory from pure mathematical speculation, the possibility of confirming or rejecting a theory's predictions through meticulous observations. Unfortunately, during the 1950s, no one seemed to be taking Gamow's ideas of primordial nuclear fusion seriously enough to mount a consistent experimental search. The turning point came in 1964, when two groups of astrophysicists built radio antennas capable of capturing the radiation left over from decoupling, one knowingly and the other unknowingly.

In Princeton, a group led by Robert Dicke was building an antenna designed to test Gamow's (and Alpher's and Herman's) prediction and measure the temperature of the leftover radiation. Interestingly, the Princeton group was not aware of Gamow's prediction, having rederived it independently, thanks mostly to the work of a young theorist named James Peebles. Not too far from Dicke's lab, Arno Penzias and Robert Wilson, working for Bell Telephone Laboratories, were searching for radio signals from a supernova remnant ten thousand light-years from Earth. Given the large distance from the source, the signals Penzias and Wilson could hope to capture were very weak. It was thus crucial that all possible sources of interference be eliminated or taken into account. But try as they might, a persistent background noise, similar to an annoying hiss in a radio, would not go away. They were even careful enough to shoo away a couple of pigeons that had happily nested inside their horn-shaped antenna. After they cleaned up the "dielectric substance" left in abundance by the pigeons (who said research is always glamorous?), the noise was still there. Even worse, it seemed to be coming from all directions in the sky; no matter where they pointed the antenna, the hissing was still there.

Flustered, they contacted B. F. Burke, a radio astronomer from MIT, who told them about Dicke's research at Princeton. The mystery of the hissing noise was solved, ironically not by the group that was looking for it; Penzias and Wilson were detecting the photons left over from decoupling, which became known, for lack of a more inspired name, as the *cosmic microwave background radiation*. "Background" since the photons are equally (homogeneously) spread throughout the cosmos, with a temperature of 2.726 degrees Kelvin (according to present-day measurements). Remarkably, recent observations showed that this temperature is the same at different points in the sky to one part in one hundred thousand! That is, if you point an antenna somewhere in the sky, measure the temperature of the radiation, and then repeat the operation somewhere else, the two temperatures will agree to 1/100,000 of a degree. "Microwave" indicates that the measured photon distribution peaks at frequencies typical of microwaves, about ten thousand megahertz (wavelengths of a few centimeters).[4]

The discovery by Penzias and Wilson marked a turning point in the history of cosmology; it put an end to the long battle between the steady-state and the big bang models, at least in the minds of most cosmologists; Hoyle (until his death in 2001) and a few collaborators still insist that it is possible to explain the microwave background within a steady-state model. However, their explanations are extremely contrived and unnatural, starting with the cornerstone assumption of the steady-state model, the continuous creation of matter out of nothing, which, to this cosmologist, is even more outrageous than a universe with a history.

A Short Discourse on Time: Part 1

The discovery of the cosmological expansion and its most direct consequence, a universe with a beginning, forced scientists to confront the issue of time and its passage. The debate over the nature of time was not new to science, having originated in ancient Greece with the pre-Socratic philosophers, who basically split into two camps: those, like Heraclitus of Ephesus, who defended the central role of transformation in nature—the cosmos of becoming—and those, like Parmenides of Elea, who posited the inherent immutability of what is truly fundamental—the cosmos of being. A compromise of sorts was achieved by the atomists Leucippus and Democritus, who held that the fundamental material entities of the world, the atoms, were eternal and indestructible, but that they combined and bound to promote the changes we see in nature. Aristotle proposed another compromise, with his division of the cosmos into two realms, the sublunar world of change and material transformation, and the supralunar world of ethereal, unchangeable objects.

Saint Augustine gave much thought to the question of time. In particular, he worried about how to reconcile time and the Creation. And don't we still, even if dressed in the robes of science? After all, a universe with a beginning raises the question of how it came into being; if not through the action of an all-pervading, all-powerful God, then how? Can science, a human invention that seeks a logical order in the patterns and cycles we observe in nature, answer such a question? The crucial difference between a religious and a scientific approach to the question of the origin of everything is that religion answers it from without, through the supernatural action of a God or gods, while science must answer it as it answers every other question, from within, as the consequence of a chain of natural causes. The creation of the universe is, for religion and science alike, a question of the first cause, the cause that triggered existence. Aristotle called it the *Primum Mobile,* the Unmoved Mover, the Being located at the outermost sphere of the cosmos, from whom all motion originates. Cosmologists euphemistically prefer to call it a problem of "initial conditions," and much recent work in cosmology deals with ways of trying to bypass such a dilemma. Augustine proposed that time appeared with the Creation, arguing that to ask what was before the Creation is nonsense because there was no "before" before time existed:

> My answer to those who ask "What was God doing before he made heaven and earth?" is not "He was preparing Hell for people who pry into mysteries." This frivolous retort has been made before now, so we are told, in order to evade the point of the question. . . . [God is] the Maker of all time. If, then, there was any time before [God] made heaven and earth, how can anyone say that [God was] idle? [God] must have made that time, for time could not elapse before [God] made it.[5]

The absence of time before the big bang is accepted by cosmology as well. The big bang has a lot in common with the space-time singularities we have seen exist inside black holes; it also marks the breakdown of our classical description of space and time by means of the laws of general relativity. If we play the cosmic movie backward, from today to the earliest moments of the universe's existence, we find matter compressed to enormous densities, as during the final stages of stellar collapse. The same quantum behavior we saw to be important near a black hole singularity will be important near the cosmological singularity; our notions of time and space simply disintegrate into a quantum foam of possible coexisting times and spatial geometries, a soup of tangled histories that go everywhere and nowhere. Thus, we also cannot talk about time "before" the big bang, because there is no flow of time to talk about. How the universe may have possibly transitioned from a primordial quantum foam of geometries into a universe endowed with a classical flow of time goes back to the determination of the "initial conditions" that set the cosmos in motion. The key idea is to combine general relativity and quantum mechanics, and try to make sense of a universe that incorporates properties akin to atoms and nuclei. We will have occasion to briefly mention some of these ideas below. Before we move on, though, we should summarize what we have learned so far, at least in regard to time.

Relativity added plasticity to the classical notion of time, posited by Newton to march inexorably forward, always at the same rate, irrespective of who is measuring it and of where it is being measured. If two observers moving with respect to each other measure an event occurring in time, they will not agree on its duration, the amount of disagreement growing with their relative velocity; fast-moving clocks tick slower. They will also not agree if one of them is in a stronger gravitational field; clocks in strong gravitational fields also tick slower.

DIAGRAM OF DISTANCE SCALES

10^{-15} m - Proton

10^{-10} m - Atom

1 m - Humans

1.3×10^7 m - Diameter of the Earth

1.5×10^{11} m -
Earth-Sun distance
(8.3 lights-minutes)

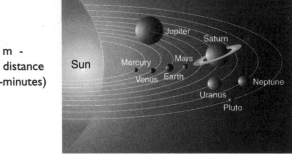

FIGURE 21: *Diagram of distances, from a proton to the observable universe.*

DIAGRAM OF DISTANCE SCALES

4×10^{16}m - Distance to nearest star (alpha-centaurii (4.3 light years)

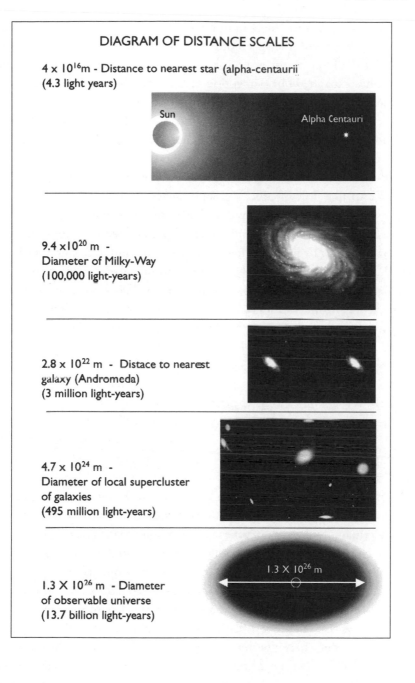

9.4×10^{20} m - Diameter of Milky-Way (100,000 light-years)

2.8×10^{22} m - Distace to nearest galaxy (Andromeda) (3 million light-years)

4.7×10^{24} m - Diameter of local supercluster of galaxies (495 million light-years)

1.3×10^{26} m - Diameter of observable universe (13.7 billion light-years)

We don't notice these differences in our everyday lives, because our speeds are too small compared with the speed of light, and the variations on the gravitational field around Earth's surface cause negligible changes to the flow of time.

Science has expanded our perception of time well beyond what our senses allow; there is a host of different time flows, lurking behind our naïve sensorial notion of a steady clockwork. Relativity refined time, plucking it from the rigid Newtonian flow, so indifferent to our presence, and rendered it manifold. But this local plasticity of time is only part of the story. Perhaps paradoxically, through the discovery of an expanding universe, modern cosmology restored the idea of a cosmic flow of time, a global, universal time, indifferent to local idiosyncrasies. We thus apparently describe reality with two times: one local, which we perceive and measure and which may, in extreme circumstances, be affected by motion or by local gravitational variations; and the other global, cosmological, which flows completely oblivious to ourselves, our concerns and measurements. This indifference of cosmological time goes way beyond human scales, stretching toward extragalactic scales, and even to the scales of the largest self-gravitating structures, superclusters of galaxies. The cosmological expansion does not stretch objects bound by local forces, such as atoms, people, planets, stars, solar systems, galaxies, or clusters of galaxies; it applies to space itself in the largest of scales, the space between galaxies and clusters of galaxies, of distances over tens of millions of light-years (see figure 21).

And yet, in spite of their apparently enormous differences, the two times, local and cosmological, are really only one, ticking forward, together. It is this universality of time that links us to the rest of the cosmos. As we grow old, so does the Sun and so does our galaxy and so does the universe. The relevant scales are, of course, completely different, one hundred years being a very long life for us but not much for a hydrogen-burning star and even less to the universe. In 1856, the German physicist Hermann von Helmholtz made a very somber prediction, which some consider the gloomiest prediction in the history of science:[6] the universe is dying, running down the way we humans do. Of course, Helmholtz knew nothing of the cosmological expansion, discovered much later by Hubble. But he knew that, according to the second law of thermodynamics (the first law says that energy must be conserved), all isolated systems—systems that cannot exchange energy with other systems—become more disordered as time goes by. The flow of time, being irreversible, is deeply related to this degeneration; it sets the rhythm of decay. A quantity called entropy was devised to measure this growth of disorganization; the more disorganized a system, the higher its entropy, and the harder to extract anything useful from it.

The second law of thermodynamics can be stated in a few different ways, which are all equivalent. In its simplest version, it says something quite obvious, that heat always flows from a hot spot to a cold one. This we know from countless practical observations, but let's imagine putting an ice cube in a glass of water, which cannot exchange heat with its surroundings, a perfect thermos; we see the ice cube melting and not the water freezing. In fact, the "system" (the ice cube and the water in the glass) reaches a final temperature above the melting point of water. The ice melts, the water cools a bit, and once they reach an appropriate mean temperature, called the equilibrium temperature, nothing else happens. An imaginary observer in the water glass finds that *time stops flowing when the ice cube–water system reaches its final equilibrium state*. In other words, time flows only when systems are not in thermodynamic equilibrium. Helmholtz extrapolated this simple idea to the universe as a whole. Since there is nothing outside the universe, the universe being all there is, it can be considered a closed system. As such, it will also obey the second law and decay in time; stars will keep radiating their heat to outer space, which will get progressively hotter as the overall entropy keeps growing. Eventually, an equilibrium situation will be reached where the temperature will on the average be everywhere the same and, just as with the ice cube–water system, nothing else will happen; time will effectively stop, and the universe will die a heat death. Helmholtz predicted the end of time as an unavoidable consequence of the second law of thermodynamics.

Helmholtz's argument was a terrible blow to humankind's self-esteem. The development of science during the eighteenth and nineteenth centuries, with its rational interpretation of natural phenomena, was followed, at least in the West, by a progressive abandonment of religion. During the Middle Ages and the Renaissance, religious beliefs responded to most of the anxieties of life and, most important, of death. There was a divine purpose to our terrestrial existence, which started with the Creation through the action of God and would again end through his actions, at the judgment day. The comfort that was to be found in faith gradually gave way to a growing secularization of Western thought, reaching a crisis level with Helmholtz's prediction of the end of time, which had nothing beatific about it. The historian of religion S. G. F. Brandon expressed this worry clearly when he wrote, "For Western thinkers there can be no more urgent task than that of resolving this dilemma, and, if possible, of producing an adequate philosophy of history, *i.e.*, of the meaning of man's life in time."[7] Since science had much to do with bringing on this crisis, I think it would be extremely irresponsible of us to turn our backs to this issue and just keep plugging away at our theories and speculations. Although many scientists have a deep aversion to philosophical issues, we would do a disservice to society if we neglected to clean up our own mess, so to speak.

The question then is . . . How? Some of my colleagues resolve this difficulty by separating science from faith and opting for a particular system of belief, which offers them solace in places where science can't go. They claim that their science illustrates even more clearly the beauty of the Creation and the wonderful spirit of God (or gods) that permeates all things. Powerful and inspiring as this compromise is, I still believe that science can do better than just offer a rational explanation of the world, which is merely reconciled with religion in the privacy of people's minds. I believe that science, in an effort to understand the unknown, transcends its more immediate role of quantifying the workings of nature. Perhaps this is what Einstein referred to as his "cosmic religious feeling," the essentially religious inspiration behind the act of rationally understanding the world. Science and religion spring from the same anxieties that baffle the human spirit. And the one common thread tying them together is our finite existence in an apparently infinite cosmos. It is in the passage of cosmic time, the same for a bacterium, a person, a star, and the universe, that we find the true unity of all things.

Time connects our existence to everything else in the cosmos, including the cosmos itself. Think that every time your heart beats, countless insects hatch out of their eggs while countless others die, waves break on all the beaches of the world, galaxies move farther away from each other, and stars are born and die somewhere in the universe. Time sets the rhythm of existence. It is when we focus exclusively on trying to make sense of the beginning and end of time, forgetting our links to existence now, that we lose our way and fall prey to either religious fervor, existentialist despair, or scientific pontification. As I have argued throughout this book, we all have a deep need to understand our origin and demise, and this need is expressed in many different ways. The mistake we make, which, I believe, is in large part responsible for the "wars" between religious belief and science and between humanists and scientists, is to forget that these different narratives inspire and complement each other and are not mutually exclusive. Time should not be divisive but inclusive. This complementary approach to the question of time could be called chronosophy: we should certainly worry about the beginning and the end, and try as much as possible to understand these events through science, faith, and art. But we should never forget that we live in between these two all-important events, and that our very existence is the product of many creations and destructions, cosmic and emotional. To paraphrase John Lennon, life is what happens while you are worrying about death. We may all be inspired by death in different degrees, but we should never be paralyzed by it.

CHAPTER 8

Time Regained

> *Worlds on worlds are rolling ever*
> *From creation to decay,*
> *Like the bubbles on a river*
> *Sparkling, bursting, borne away.*

> — PERCY BYSSHE SHELLEY,
> *Hellas*, LINES 197–200[1]

The third main stage of development of modern cosmology is best characterized by a double explosion—of theory and observation. On the theoretical side,
the triumph of the big bang model led to an increased effort to understand the
earliest stages of the universe's history, made possible through an influx of ideas
from high-energy particle physics. If nucleosynthesis—the synthesis of the
lightest nuclei between one second and three minutes after the bang—was
understood through the application of nuclear physics to cosmology, for earlier
times, we must use a description of matter at energies higher than nuclear, when
it is broken into its most fundamental constituents. This is where my research
comes into the game, in trying to make sense of the cosmos in its earliest infancy.

On the observational side, the results of Penzias and Wilson were further
confirmed and refined to an amazing degree by a series of spectacular measurements of the properties of the cosmic microwave background. These were
achieved through Earth-based experiments, such as microwave detectors flown
over the South Pole on high-altitude balloons, and through the Cosmic Background Explorer (COBE) satellite, which measured the microwave background over the whole sky and constructed a map with an angular accuracy of
about ten degrees. This map is a snapshot of the universe when it was only
300,000 years old. Like any map, it invites us to travel to its lands with their

promise of wonders and riches. And many riches were indeed found, and remain to be found in the next few years, as new satellite missions produce much more detailed maps. For although the COBE map confirmed the amazing temperature homogeneity of the cosmic microwave background, it also revealed what lies beneath this calm ocean of radiation: small temperature fluctuations that carry information from the very early moments of cosmic history (see figure 35 in insert).

But the microwave background is only one aspect of a vast observational effort, which involves, among other things, telescopes scanning all wavelengths, from radio waves to gamma rays, allowing us to study objects that are several billion light-years away, and thus older than the solar system itself. These observations have revealed a very complex universe, where galaxies cluster in long three-dimensional filamentary webs that can cross tens of millions of light-years, while at other spots there are enormous regions practically without any matter, the cosmic voids (see figure 36 in insert). An image often used to help visualize the large-scale structure of the cosmos is that of a bubble bath, with soap-bubble walls touching and coalescing to form a rich froth. Now, sprinkle black pepper on the froth (without popping the bubbles), and observe the distribution of the pepper grains. If you imagine each pepper grain as a galaxy, you get a rough picture of the universe at very large scales: grains cohering around empty space. As we will see, modern cosmological theory tries to weave these two key observational discoveries together, showing that the temperature fluctuations of the microwave background are tied in with the fluctuations that generated, through gravitational instabilities, the rich and complex large-scale structure of the observable universe. What's more, the theory predicts that these fluctuations were generated at the universe's earliest moments of existence, the physics of the very small, elementary particles and their interactions, influencing the physics of the very large, the past tied to the present and to the future.

The Cosmos, circa 1998:
Building Blocks

I will start by briefly reviewing some of the ideas of particle physics from the last three decades. The goal of high-energy particle physics is to find the smallest constituents of matter, the basic building blocks of everything in the cosmos. This mission, of course, assumes that these constituents exist; if you keep dividing matter into smaller chunks, you will eventually reach the smallest chunks, the elementary particles. During the 1940s and 1950s, experiments involving high-speed collisions between heavy nuclei and smaller particles, such as electrons and protons, revealed an extremely rich subatomic world, populated by *hundreds* of other particles of matter. This proliferation of particles was a direct consequence of Einstein's prediction that matter and energy are interconvertible, given the proper conditions. Imagine colliding an electron and its antiparticle, the positron, which we encountered in chapter 5. Their identical masses contribute to the total energy of the electron-positron system before they collide. But they also have kinetic energy, energy of motion, which grows with their velocity. Thus, the total energy of the electron-positron pair before they collide is much higher than their rest masses, because of the contribution from the energy of their motion. When they collide, as we have seen, they disintegrate into photons, which carry all the energy of the electron-positron pair, their rest mass and their kinetic energy. The photons may then create other particles, which can be much more massive than the original electron-positron pair. The only requirement is that the total energy before and after is the same. There is thus a wonderful transmutation of energy into matter, following roughly a relation like this: electron + positron + lots of kinetic energy → highly energetic photons → heavier particles + some kinetic energy. A physicist once remarked that an equivalent phenomenon at human scales would be to collide two tennis balls and get two elephants!

This proliferation of particles was clearly a major challenge for those who searched for the basic constituents of matter; after all, what is the meaning of elementary building blocks if you have hundreds of them? The resolution of this dilemma was proposed in 1963 by the Caltech physicist Murray Gell-Mann, and it bears some resemblance to how we explain the periodic table of elements. There are ninety-two naturally occurring chemical elements, all made of the same three basic constituents, electrons, protons, and neutrons. By combining these three particles in different numbers, we recover all the ninety-two chemical elements. Gell-Mann proposed that a similar idea could explain the hundreds of particles that appeared in the high-energy collision experiments. They all had one very important property in common; they interacted via the strong nuclear force, the same force that binds neutrons and protons together in the atomic nucleus. Particles that interact via the strong force are called *hadrons*. Gell-Mann showed that all hadrons can be described as combinations of more fundamental particles, which he named quarks, taking a word from James Joyce's *Finnegans Wake*. We now know that there are six fundamental quarks in nature, which go by the names up, down, charm, strange, beauty, and top. The proton, for example, is made of three quarks, two ups and one down (uud), while the neutron is made of two downs and one up (udd). Present particle experiments give no indication that any additional quarks exist.

Apart from the six quarks, there is another group of six particles known as leptons, from the Greek word for "lightweight." We encountered two of them, the electron and the electron neutrino, which can in fact be thought of as forming a pair. The other four leptons are also grouped in pairs, the muon and its neutrino, and the tau and its neutrino. The muon and the tau are pretty much like heavy electrons, the muon being about two hundreds times as massive, while the tau is almost four thousand times as massive. Indeed, the tau is almost twice as heavy as the proton, making the name lepton, "lightweight," an interesting misnomer. These twelve elementary particles, the six quarks and the six leptons, are elegantly grouped into three "families" of four members each. (To each particle we can add its antiparticle as well.)

THE THREE FAMILIES OF ELEMENTARY PARTICLES

electron	muon	tau
e-neutrino	m-neutrino	t-neutrino
up	charm	bottom
down	strange	top

TABLE 2: *The three families of elementary particles.*

The Cosmos, circa 1998:
The Quartet of Fundamental Forces

The matter we are familiar with—protons, neutrons, and electrons—belongs exclusively to the first family. Members of the other families appear only in highly energetic astrophysical phenomena or in particle collisions on Earth. We thus arrive at a classification of matter in terms of only twelve elementary particles, six quarks and six leptons (and their antiparticles). But this is only half of the story. In order to describe the physics of matter, we must understand how those building blocks interact with each other. After all, we cannot build a sturdy house simply by piling up bricks on top of each other; we need mortar. Here enters the concept of the fundamental forces of nature, which we believe are four in all. We have encountered three of them so far: the *gravitational force*, the *electromagnetic force*, which acts on any particle with an electric charge, and the *strong nuclear force*, which binds quarks together into hadrons, such as protons and neutrons, and also binds protons and neutrons into atomic nuclei. The last member of the quartet of fundamental forces is the feeble *weak nuclear force*, whose most important effect is to promote the radioactive decay of unstable nuclei, such as uranium or plutonium. Like the strong nuclear force, the weak force acts only within nuclear distances; in our everyday lives we experience only the gravitational and electromagnetic forces, the ones capable of being felt at long range.

So, the subatomic world is described in terms of twelve particles of matter and four fundamental forces. These fundamental forces are also described in terms of particles, messengers carrying the information about the interactions between the different matter particles. Each force has its own set of carriers, which we call the particles of force. For example, electromagnetic interactions, such as the electric repulsion between two electrons, are represented by an exchange of photons. A suggestive image often used is that of two ice-skaters throwing baseballs at each other; the energy and momentum of the baseball is exchanged between the two skaters, establishing an interaction between them. The photons are the carriers of the electromagnetic interactions. Note that photons are massless particles, their energy being measured only by their frequency. This "masslessness" is no coincidence, but a consequence of the fact that the electromagnetic force is long-range, decaying with the square of the distance from its source, vanishing only when the source is infinitely distant. If the photon had a mass, the electromagnetic force would be short-range, like the weak and strong nuclear forces. The same is true of gravity; since the gravitational attraction between two bodies also decays with the square of their separation, the *graviton*, the particle conjectured (it has not been observed yet) to transmit the gravitational force, is also massless.

Since the strong and weak forces are short-range, their description in terms of carriers must be different. The weak nuclear force has not one but three force carriers, which were in the 1960s predicted to exist by Sheldon Glashow, Abdus Salam, and Steven Weinberg, all with masses within eighty to ninety times that of a proton. These three particles, which go by the uninspired names of W^+, W^-, and Z^0 (the superscripts denote their electric charges), were found in the early 1980s by the Italian physicist Carlo Rubbia and his team at the European Center for Particle Physics (CERN). This remarkable triumph of theoretical and experimental particle physics was recognized with well-deserved

Nobel Prizes and shaped the dreams of a whole generation of young physicists growing up in the 1970s and 1980s, including this one. What could be more exciting than contributing to the grand mission of unveiling the innermost constitution of the material world?

I have met the three theorists at different stages of my career. Weinberg's *The First Three Minutes*, an early popular account of big bang cosmology and the discovery of the cosmic microwave background, greatly inspired me to become a cosmologist, especially one who applied particle physics to the early universe, as he had masterfully done. I met him in the mid-1980s, during a meeting of the Royal Society in London, when I was a graduate student at King's College. He certainly does not remember this, but I sat by his side, completely transfixed by being close to such a great man; Weinberg's textbook on general relativity and cosmology was (and still is) a kind of bible to me. What I most remember is the profusion of nervous energy coming out of him; his leg bounced nonstop, he dropped his newspaper quite a few times (which I offered to pick up with a pathetic reverent look in my eyes), and continually looked around as if searching for someone or something. Perhaps he was just checking his audience. Although I really wanted to tell him about my thesis work on cosmology in higher dimensions (a popular topic then and, curiously, now again), his attitude, mixed with my own shyness, was not very inviting. I sat in silence and paid attention to what he had to say during his talk.

Abdus Salam knew of my work when I was still a third-year graduate student, because he was also interested in the idea that the universe perhaps has more spatial dimensions than the familiar three. The key question at the time was to try to make sense of how this could be; that is, how could a universe, which started with ten or eleven dimensions, be so obviously four-dimensional today. (These numbers include one dimension for time.) My work, together with my Ph.D. adviser, John G. Taylor, was to dream up scenarios where the "extra" spatial dimensions got curled up into a ball, which somehow did not expand with the rest of the universe. Thus, as time went by, there emerged a clear separation between the three-dimensional space we live in and the extra, "compactified" dimensions, which remained much, much smaller. Why were we interested in a universe with so many dimensions? The idea was that in a higher-dimensional universe it might be possible to interpret all four forces of nature as being initially one force, the "unified field." As the universe expanded, its geometry split, causing the unified force to differentiate into the four forces we see today.

This apparently strange approach was first put forward in 1919 by the Polish mathematician Theodor Kaluza, who showed that in a five-dimensional universe, electromagnetism, the only other force known then, could, like gravity, be interpreted geometrically. What we perceive as a four-dimensional world with gravity and electromagnetism can be explained as a five-dimensional world with only one force. The distinction between the two forces comes from a peculiar choice for the geometry of this five-dimensional world, which can be best visualized as cylindrical, like a stretched garden hose; imagine that the four usual dimensions we live in (three space and one time) are represented by the (infinitely) long axis of a "cylinder." Each point along the axis has a circle "attached" to it, the cross section of the cylinder, as illustrated in figure 22. This circle is associated with the extra, fifth dimension, which is thus perpendicular to our four-dimensional reality. Just as a hose, when seen from afar, looks like a one-dimensional line, the extra circular dimension can be "seen" only at extremely small distances. In 1926, the Swedish mathematician Oskar Klein applied quantum mechanics to Kaluza's five-dimensional unification scheme to show that the radius of the extra dimension could be as small as the smallest length scale where our description of gravity by means of general relativity still makes sense, called the Planck length, equal to 10^{-43} centimeters. It is no wonder that, if the universe does indeed have extra spatial dimensions, they have so far eluded direct detection; present high-energy experiments probe distances of about 10^{-16} centimeters. According to the Kaluza-Klein theory, what we perceive as electromagnetism and gravity in our world are but channeled-down vibrations of a peculiar five-dimensional world where gravity rules alone. During the 1980s, it was shown

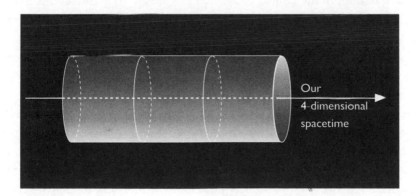

FIGURE 22: *Schematic diagram of Kaluza's five-dimensional "cylindrical" universe: a "circle" is attached to each point in our four-dimensional space-time.*

that, in principle, the Kaluza-Klein approach could be extended to include the strong and weak interactions. This being the case, the four forces of nature all originated from a single force living in a higher-dimensional space. It was an exciting time to be doing a Ph.D. in high-energy physics.

For my thesis, I had obtained several solutions of Einstein's equations in these higher-dimensional worlds, investigating in particular their cosmology, that is, how these geometries would evolve in time. This approach was reminiscent of the desktop universes of the first half of the twentieth century, for we had no observational evidence (and still don't) to guide our mathematical models, only physical intuition and consistency; we know what properties these theories must have when reduced to four dimensions, since they must conform to the universe we live in. These were higher-dimensional desktop universes, generalizing Kaluza's old idea to include all four forces of nature into a single geometrical scheme, a grand realization of Plato's dream of an underlying reality of perfect geometrical forms translated into modern physical thought. Had we succeeded, we could have proudly stated, "In the beginning all was geometry." But we haven't yet.[2]

Salam was extremely generous to me, quoting my work in his talks and papers, much to my amazement; imagine that—a Nobel Prize winner citing the work of a Brazilian graduate student. This attention to my work, simple as it was, gave me tremendous self-confidence. I think Salam knew that and, as head of the International Center for Theoretical Physics in Trieste, Italy, wanted to help young physicists from Third World and developing countries to succeed. My adviser invited him to be the external examiner (who at first sounds more like the chief inquisitor) at my Ph.D. thesis defense, which, in England, but not in the United States, is a private affair, involving just the student, the adviser, and the external examiner between four walls. For about an hour, Salam asked me the usual questions about my work, mostly focused on the motivations for assuming this or that behavior, and all went well enough. Inspired by all this, I asked him to write me a recommendation letter for a postdoctoral research position, a two- to three-year appointment that is a much needed stepping-stone toward the elusive university professorship. I ended up accepting a postdoctoral position with the astrophysics group at Fermi National Accelerator Laboratory (Fermilab, for short), near Chicago, which was then, and still is, a powerhouse of ideas in cosmology and astrophysics. It was at Fermilab that I met my longtime collaborator and friend Rocky Kolb, who became my true mentor during my transition to American academia.

My encounter with Sheldon Glashow was of a more "domestic" nature.

Terry Walker, my Fermilab colleague now at Ohio State University, had gotten a postdoctoral position at Boston University and invited me to give a seminar there. The arrangements surprised me a bit; instead of the usual motel room, I was to spend a week as a guest at Glashow's home.

It was quite a week, also thanks to my Brazilian friend Angela Olinto, who was finishing her Ph.D. at MIT, and knew where to go and what to do in Boston. I was invited to Shelly's birthday celebration, which was clearly not his favorite time of the year, a small affair including two other professors from BU, animal masks brought by one of them, and a string quartet. The highlight of the week, apart from my seminar and my outings with Angela and her friends, came on Saturday morning, when Shelly offered to make me cappuccino in his kitchen, and we spent quite some time talking about physics and movies. We both had watched *Dark Eyes* the night before, the romantic Italian-Russian movie starring Marcello Mastroianni, which Shelly clearly enjoyed, judging by his big smile on the way out from the theater. "Wasn't that wonderful?" he said, his eyes glowing. It may have been the elegance and poetry of the photography, or the impossibility of the love affair between the married Mastroianni and an elusive Russian woman. In any case, it was obvious that we felt something in common, perhaps a longing for the intangible, hard to describe but easy to recognize.

The Quest for Unification: The Inner-Space/Outer-Space Connection

Back to the four fundamental forces. The weak force is thus carried by the three heavy particles postulated by the Glashow-Salam-Weinberg models and found at CERN. What about the strong nuclear force? We have seen that protons, neutrons, and all the hadrons, the particles that interact via the strong force, are made of quarks and their antiparticles, the antiquarks. The particles that keep the quarks together in the hadrons, acting as a kind of nuclear glue, are called, properly enough, gluons. There are eight gluons, all massless like the photon. That the strong force has massless carriers and still is a short-range force, an apparent contradiction, is due to a somewhat mysterious property known as *confinement;* the best way to visualize what goes on inside a proton is to think of a bag with three quarks somehow stuck inside, as if the gluons were elastic strings connecting them. The amazing property of these strings is that the farther you try to separate the quarks, the harder the strings pull them back together; in fact, to pull the quarks outside a proton, you have to spend so much energy that the string snaps and two new quarks appear at the broken ends. In other words, the quarks are confined inside the hadrons in such a way as to make it impossible to liberate them. We cannot see free quarks roaming around the world. Only through highly energetic particle interactions can we conclude that protons, neutrons, and other hadrons have an inner structure represented by three point-like particles inside a small region permeated by a gluonic sea, the quarks in the bag.

We can think of nuclear matter as having two distinct phases, just as liquid water and ice are two distinct phases of water; at low energies, nuclear matter appears as protons, neutrons, and other hadronic particles, with no sign of its inner quark-gluon structure. But at high enough temperatures (and thus energies), equivalent to about one trillion degrees Kelvin (hundreds of times higher than at the core of a supernova), the hadrons melt, so to speak, into a quark-gluon plasma. At these incredibly high temperatures and densities the reverse of confinement happens; when squeezed close together, the quarks behave like free particles, a property aptly named asymptotic freedom. There are two places where a quark-gluon plasma can exist; before one hundred thousandth of a second after the big bang, when temperatures were sufficiently high, or at highly energetic nuclear collisions, such as those now underway (2002) at the Relativistic Heavy Ion Collider (RHIC) at the Brookhaven National Laboratory, on Long Island.

This transition of the properties of nuclear matter is crucial to our understanding of the early history of the universe. Recall that Gamow started his primordial soup with protons and neutrons, that is, hadrons. If we want to go further back in time, toward the initial cosmological singularity, we should really start the soup with quarks and leptons, interacting via the four fundamental forces, as determined by modern particle physics. This recipe would be correct at least around one millionth of a second after the bang. As we will see, at earlier times other changes occur. The lesson here is clear: we cannot study the history of the very early universe without incorporating particle physics into cosmology. And since, as we will also see, the "end" depends on the beginning, we cannot investigate the future evolution of the universe without knowing its history; in many ways, the universe's childhood determines much of its mature life, a notion that echoes our own private histories. The universe cannot run from its past.

Event	Time	Temperature
Decoupling	300,000 years	1000 Kelvin
Nucleosynthesis	1 sec	10^{10} Kelvin
Quark-hadron	10^{-5} sec	10^{13} Kelvin

Table 3: *Important cosmological events discussed so far.*

We want to keep going backward in time, getting as close as we can to the initial event. However, as we go backward, we must deal with physics at higher and higher energies; we have seen this going from Gamow's proton-dominated primordial soup, fine for describing the synthesis of light nuclei at times of roughly one second after the bang, to the quark- and lepton-dominated soup at times of a millionth of a second after the bang. The next step backward initiates a trend that will repeat itself all the way to the beginning, or pretty close to it. The trend has to do with a deep change in the way particles interact at energies higher than the ones causing the hadrons to melt into quarks; the four fundamental forces, which regulate how particles interact, start to shed their unique behaviors and, in pairs, act more as a single force. The higher the energy, the more the forces approach one another, one by one, until, at the earliest of times, there is only one force.

We see that there are, in fact, two avenues toward unification. One could be called the geometrical approach, which uses extra spatial dimensions to describe all forces in equal footing, a top down approach. This is the case of the Kaluza-Klein unification, which, for all its attractiveness, has problems including ordinary quarks and leptons; it works well for the particles of force but not for the particles of matter. A very important variation on Kaluza-Klein higher-dimensional unification involves superstring theories, where the notion of an elementary particle is replaced by the notion of fundamental strings. According to these theories, what we call elementary particles in our low-energy reality are really different vibrational modes of the fundamental strings, a truly elegant approach to unification. Unfortunately, because of their mathematical complexity, superstring theories are still very much a work in progress, although a promising one. I will restrict myself to the second avenue toward unification of forces, which remains in four dimensions, the bottom-up approach. It leaves gravity aside but attempts to bring the electromagnetic, weak, and strong forces together as energies increase. Irrespective of the role superstring theories may play at the earliest stages of the universe's history, given our focus on questions related to the universe's future, we should be able to make progress without traveling deep into extra dimensions. In fact, we will adopt what cosmologists call the "low-energy effective theory" approach, in which the information from the universe's hazy beginnings (superstrings or not) is encoded into a simple model that supposedly encapsulates most of what we need to carry on.

The Quest for Unification: Toward the Beginning

The first unification brings electromagnetism and the weak nuclear force together at temperatures of one thousand trillion (10^{15}) degrees Kelvin. At temperatures higher than this, the weak nuclear force, which we perceive from our mundane lower-energy reality as a short-range force, becomes long-range, just like electromagnetism. In practice, the three force carriers of the weak nuclear force become as massless as the photon above this temperature; the two forces behave as one with four massless carriers. This deep change in the behavior of these two fundamental forces is at the borderline of what we can study with particle accelerators here on Earth, in particular at Fermilab and CERN. We have great confidence that our description, based on the Glashow-Salam-Weinberg model, is correct, although one key ingredient is missing, the elusive particle known as the Higgs boson. The Higgs, named after the Scottish physicist Peter Higgs, plays a crucial role in electroweak unification, as the particle responsible for giving mass to all twelve (or nine, if the neutrinos are massless) massive quarks and leptons. We can think of the Higgs as a sticky fellow that hugs every particle with a given strength, known as a coupling constant. It hugs the electron with a certain strength and the up quark with another strength. Once the Higgs hugs a particle, the particle is doomed to carry it around like a permanent backpack; the net result is that the strengths of these hugs, or, if you want, the masses of these backpacks, determine the masses of the particles on a one-by-one basis. What controls the hugging behavior of the Higgs is the ambient temperature. For temperatures above the scale of electroweak unification, around 10^{15} degrees Kelvin, the Higgs is pretty much innocuous, for the heat agitation interferes with any hugging attempts. But as the temperature drops, which, as we have seen, necessarily happens as the universe expands, the Higgs begins to stick to the quarks and leptons with different strengths, giving them the masses measured at low energies.

At this point, it is worth introducing the concept of a field. Every particle in nature, be it a particle of matter, like an electron or a quark, or a particle of force, like a photon or a gluon, has a field associated with it. In fact, particle physicists prefer to describe particles as being excitations of their fields, somewhat the way a sound note on a guitar string is an excitation of the string. The string can be excited in many different ways, producing notes of a different pitch (frequency) and loudness (amplitude). Likewise, the field can be excited in many different ways, producing particles of different energy and momentum. When we apply particle physics to the early universe, it is often more convenient to describe matter in terms of continuous fields, as opposed to their associated discrete particles.

The change from a high-energy world, where the electromagnetic and weak interactions are unified into a single electroweak interaction and all particles are massless, to a low-energy world, where the two forces act separately and the quarks and leptons are massive, is also a phase transition controlled by the value of the Higgs field. Just as water has different properties at different temperatures, so do the fundamental forces. The net result of electroweak unification is that the universe, at temperatures above 10^{15} Kelvin, is a simpler place, where all particles but the Higgs are massless and their interactions are described by three, rather than four, fundamental forces. This simplicity is accompanied by a higher symmetry as well; just as in liquid water molecules can be found equally in every position and direction (a highly symmetric state) and in ice they are lodged in rigid crystalline lattices (a less symmetric state), the unified universe has a higher degree of symmetry than the low-energy universe. This symmetry is of a more abstract nature than the positional symmetry of water versus ice, having to do with mathematical properties of the particle interactions themselves, whose details are of no concern to us here. But it is symmetry nevertheless, lending elegance to a unified universe, a step closer to the Platonic ideal of mathematical perfection as nature's most fundamental property.

The electroweak unification corresponds to times of less than roughly one trillionth of a second after the bang (10^{-12} seconds). This may seem like an outrageously short timescale to us, but it is extremely long for a photon, whose cruising time across a proton is a mere 10^{-24} seconds—that is, a trillion times smaller! So, don't be alarmed as we continue to march on to even earlier times and higher temperatures, although, as we do so, we leave behind the safety of observations and move into the realm of sound theoretical speculation. As we will see, it is presently extremely hard, but not impossible, to test physics at the very high energies prevalent in the early universe.

The next step toward bottom-up unification of forces is the incorporation of the strong force into the electroweak unification. This theory, known as grand unified theory (GUT for short), was initially developed during the 1970s by Glashow and Howard Georgi of Harvard University. It remains unconfirmed, although many believe its confirmation is just a question of time. Quite a few variations on the GUT theme have appeared, different schemes to bring the strong and electroweak forces together. But they all have one property in common; since the strong interaction rules the behavior of quarks, while the leptons interact via the weak force (and both via electromagnetism), a unified theory of all three interactions should allow for quarks to transmute into leptons and vice versa. Leptons and quarks are part of our low-energy, asymmetric reality; in the unified world of GUTs, they can easily become each other. An immediate consequence of this prediction is that the proton, being made of quarks, should not be a stable particle but should decay as its quarks mutate into leptons; the matter we are made of is not eternal. The predicted lifetime of protons has changed over the years, but it started close to 10^{30} years, apparently a ridiculously long span, compared with the universe's age of about fourteen billion years (1.4×10^{10} years). However, if we fill a big enough tank with water, there will be plenty of protons for us to witness the occasional decay. This experiment has

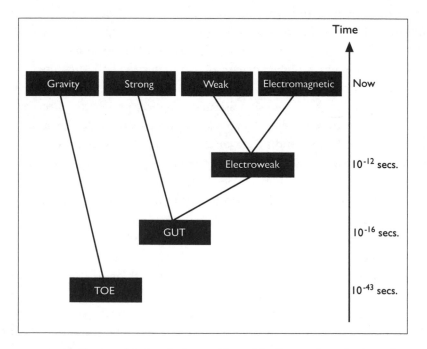

FIGURE 23: *Unification of the four fundamental forces. The diagram shows the times when unification of the different forces is expected to occur during the history of the universe. TOE stands for "Theory of Everything." (Times are approximate.)*

been attempted at different underground tanks (underground to avoid spurious decay signals) across the world, but the proton has stubbornly refused to decay. Learning from these experiments, theorists rushed to modify their GUT models, allowing for longer-lived protons. So, even though the proton's longevity ruled out the simpler models of GUTs, there are many variants that have protons sufficiently long-lived to elude the water tank measurements.

Grand unification works in somewhat the way electroweak unification does; at temperatures above a certain value (the enormous 10^{27} degrees Kelvin), the universe is best described as having two forces: gravity and the grand unified force. This GUT force is also carried by particles, at least twenty-four of them (in its simplest version). There is also a hugging Higgs, which determines the two phases of the theory; at high temperatures every particle is massless and the GUT force is long-range, while at low temperatures this GUT-Higgs gives masses to every particle that needs one. As the universe expands and cools below the critical temperature of GUT unification, the grand unified force splits into the strong and electroweak forces and the initial large symmetry is reduced. The actual mechanism by which this reduction (or breaking) of symmetry occurs is still very much a matter of contention, not just for grand unification but also for electroweak unification. But this is the subject of a whole other book.

The Cosmos, circa 1998:
Cosmic Challenges

It does seem that we have it all figured out, a hot big bang model married to the unification of the forces, an elegant description of the universe consistent with the motto "The earlier the simpler." But problems and open questions abound. On the particle physics side, we don't know, for example, why protons and electrons have the same (and opposite) electric charge, or why their masses are what they are. As we build our models to describe their interactions, we use the values measured in the laboratory as input parameters. Clearly, a much more satisfying theory would be able to *predict* these values from a fundamental level of description. We have but a hazy understanding of nature, as if its true reality, in all its stark and profound beauty, were covered by a veil. This is what physicists working on superstring models are trying to do, lift the veil and pierce right through the essence of physical reality. Only time will tell whether reality is indeed fundamentally described by vibrating strings.

Even with the influx of ideas from particle physics, big bang cosmology is far from complete. Here, too, we must use several input parameters, which we obtain from observations and feed into our models to obtain a consistent picture of the universe, but would like to understand from more fundamental physical processes. One of them is the curvature of the universe. As we have seen, whether the geometry of the cosmos is flat, open, or closed depends on two key contributions to Einstein's equations, the total energy density of matter and radiation and the cosmological constant or some similar sort of "antigravity" component. Since the fate of the universe is determined by its geometry, under-standing what makes the cosmos bend is deeply related to understanding its destiny. There are two schools of thought among cosmologists. The *purists* believe that the curvature of the universe was determined during its early moments of existence. Furthermore, their calculations and models predict that *the universe is flat.* To the purists, it is just a matter of time before the observers amass enough evidence to confirm their prediction. The *gazers* believe that we should measure and find out, that the models upon which the purists base their prediction for a flat universe are too speculative to be taken as the final call. Let us delve deeper into this controversy.

Alan Guth, now at MIT, originally developed the theory that *predicts* the flatness of the universe and goes by the name of inflation. Ideas hinting at Guth's solution were floating in the air by the late 1970s, but no one had applied them yet to the relevant context or with Guth's disarmingly simple clarity and elegance. His paper appeared in 1981 and was quickly followed by variations by Andrei Linde (now at Stanford) and independently by Andreas Albrecht (now at the University of California at Davis) and Paul Steinhardt (now at Prince-ton). Since the appearance of Guth's seminal paper, there have been dozens of scenarios—some admittedly by this author—assuming different recipes for the primordial soup of particles, but achieving essentially the same final results, with better or worse tweaking of the parameters. It is only fair to say that infla-tion is still very much an idea in search of a theory; much of the present debate as to whether this or that model is the best or "most natural" boils down largely to aesthetical considerations, a dangerous criterion when applied almost by itself. After all, beauty is in the eye of the beholder. In any case, the general idea of inflation is so simple and elegant that most of us expect it to survive in some form or another whenever our imagination graces us with a compelling theory well motivated by fundamental physics and consistent with observations. Or, at least, a theory that can be clearly distinguished from other competing theories through observations.

Before we discuss how inflation predicts the flatness of the universe (in most of its versions anyway—some are consistent with an open geometry), it is worth investigating another of its triumphs, the resolution of the *horizon problem*. One of the most glaring limitations of the standard model of cosmology, the hot big bang model, is its failure to explain its best-measured property, the amazing homogeneity of the microwave background temperature. That an observer can point her microwave antenna anywhere in the sky, Northern or Southern Hemisphere, and measure the same temperature to an accuracy of one part in one hundred thousand is quite remarkable. But this temperature democracy becomes even more remarkable, actually puzzling, when we try to understand how that can be so. As we have seen, Einstein's special theory of relativity relies on the principle that the fastest speed with which information can travel is the speed of light. This includes how fast different regions of the universe can communicate—through interactions of particles with photons—to regulate their temperature. Because of the finiteness of the speed of light, a region where causal contact is allowed, the causal horizon, surrounds each point in the universe (say, Earth). As the universe becomes older, the boundary of the horizon distance around each point grows at the speed of light. Thus, any region beyond the horizon is also beyond reach; we cannot possibly know what goes on there. For example, the Sun has existed for about five billion years. An observer on a planet six billion light-years away from the Sun will know of its existence only in one billion years. Clearly, particles belonging to regions outside each other's causal horizons cannot exchange information. Let's illustrate this point with a more prosaic example.

Prepare a very hot bath and measure its temperature at different spots. After being satisfied that the water temperature is fairly homogeneous, throw in a bucket of very cold water at one end of the bathtub; clearly, it will be some time before the temperature at the other extreme of the bathtub will change as a result of the incoming cold water. Microscopically, the hot (fast) water molecules will collide and exchange energy and momentum with the cold (slow) molecules, until temperature inequalities are mostly evened out. Of course, collisions between molecules do not stop once the temperature is everywhere the same, but their average velocity will remain nearly constant. The point is that it takes time for differences in temperature to be equalized. You can also test this, but carefully, by placing a metal poker into a hot fireplace. Heat will propagate upward toward the end of the poker until it reaches your hand. This equalization of temperature, be it in the bathtub, the hot poker, or the universe, happens because of interactions between the material components in each of these mediums. Water molecules collide more or less in response to temperature changes, as do atomic and structural vibrations of the metal poker, and as do photons and other particles in the early universe.

The puzzling fact for the big bang model is this: as we have seen, the last time photons interacted with matter and could thus adjust their temperatures was at decoupling, when the universe was 300,000 years old. The causal horizon at that time—the region within which causal processes moving at most with the speed of light could have smoothed out temperature fluctuations—corresponds to an area in today's sky that is smaller than one degree (about twice the size of a full Moon). This being the case, how can photons across distant regions of the universe "know" they should all be at the same temperature? Here is another way of visualizing this: imagine that you cover a large transparent globe with quarters. Now transport yourself to the center of the globe. Each quarter represents (not to scale!) the size of the causal horizon at decoupling, while the whole globe represents the causal horizon today. You could also imagine pointing an (infrared, if quarters are at ambient temperature) antenna at different quarters and measuring their "temperature." There is no a priori reason for all the quarters to have the same temperature, unless they were all heated in the same way. This, in a nutshell, is the horizon problem.

Guth proposed the solution. Suppose that very early on in the history of the universe, at around the same time a GUT is plausible (at 10^{-36} seconds), a very abrupt change occurred to the expansion rate, causing a tiny, causally connected region to be violently stretched to an enormous size, so huge as to comprise all of our observable universe today, as is illustrated in figure 24. After a very brief period, the ultrafast expansion rate slowed down and the now "inflated" universe recovered the usual expansion rate of the standard big bang model. (Recall that in the usual big bang model gravity gradually, but consistently, slows down the expansion.) Because of this extremely short-lived and fast-paced period of expansion (for the mathematically savvy reader, exponentially fast), Guth's proposal and all its variations became known as inflationary cosmologies.[3]

The Cosmos, circa 1998: Inflating the Big Bang

How does this short period of accelerated expansion solve the horizon problem of the standard big bang model? Consider again the globe covered with quarters. According to the standard scenario, the photons in each of the quarters would have nothing to do with the photons in the other quarters; the fact that their overall temperature is so remarkably homogeneous would be a mystery. Within the inflationary scenario, there wouldn't be separate causally discon-

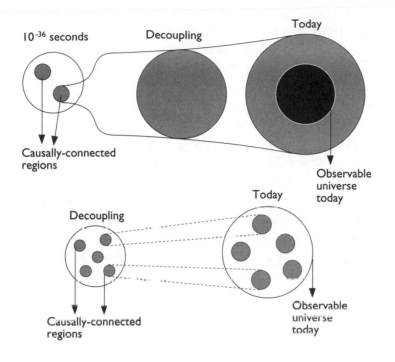

FIGURE 24: *How inflation solves the "horizon problem." In an inflating cosmology* (top) *the whole observable universe fits well within one causally connected patch; in a standard Friedmann cosmology, the observable universe today contains many regions that, at decoupling, were causally disconnected. (Diagram not to scale.)*

nected quarters patching up our observed universe, but the whole sky would have originated from a single connected region, as if one coin covered the entire globe.

The diagrams in figure 24 attempt to illustrate the differences between an inflating and a noninflating cosmos. The key point is that, in an inflating cosmos, the photons that we measure today all come from (or are the progeny of processes within) the same causally connected region, thus ensuring their equal thermal properties. A noninflating cosmos is like a causally disconnected quilt, each patch having its share of photons with properties unrelated to those in neighboring patches.

Apart from offering a plausible solution to the horizon problem, inflationary cosmologies also solve a host of other shortcomings of the big bang model. One that we briefly mentioned above is the flatness of the universe. At the beginning of this chapter, we saw how the global geometry of the universe is determined by the contributions to its energy density from matter, radiation, and a possible cosmological constant or something that mimics its effects. During the early 1990s, a host of different observations placed the energy density of the universe to within 30 percent and twice the critical density, that is, close to flatness but still undetermined. The observational data at the time provided no compelling reason to include a cosmological constant in the game, although some models with it were proposed to stave off a reincarnation of Hubble's age problem: measurements of the oldest stars in small galaxies called globular clusters indicated they were older than the universe, according to *some* measurements of its age. But neither the stellar age measurements nor the age of the universe measurements were conclusive enough to justify the inclusion of a cosmological constant; there was room to have a cosmology without it. (One of the key players who helped dispel this "age problem" was my Dartmouth colleague Brian Chaboyer. He was one of the good guys who made the stars younger than the universe.) The (fairly broad) consensus by late 1997 was that, with no cosmological constant, the geometry and destiny of the universe should be completely determined by the amount of matter in it.

Easy, right? All we had to do was measure how much matter was out there, compare it with the critical density, and get the answer. But here trouble begins. First, theory alone indicated that an old universe such as ours must be pretty much flat or it would not have aged as it has. If it had been overdense, gravity would have caused it to collapse a long time ago into a big crunch; if it had been underdense, it would have expanded too fast for galaxies and thus stars and life to develop. The mere existence of an old universe therefore points toward its flatness. This is known as the *flatness problem* of the standard big bang cosmology. Something must have fine-tuned the initial balance between gravity and expansion with incredible precision in order for the universe to still be here, filled with galaxies and people. Think of how precise a target shooter must be to hit the bull's-eye; a bit up she overshoots, a bit down she undershoots. So it is with the universe.

But the observations were telling something quite different—that there was not enough matter out there to make the universe flat. The matter that makes up things that shine, such as stars, was estimated to compose only about 1 percent of the critical amount for flatness. To that, we should add matter that doesn't shine by itself, like stars too light to ignite full-blown thermonuclear reactions (called brown dwarfs), planets, asteroids, and cool gas clouds. This matter is part of what is called the dark-matter component of the universe: we can infer its existence only by the gravitational effect it has on matter that shines. Still, ordinary dark matter—made up of protons, neutrons, and electrons—did not help much, adding up to at most 5 percent of the critical density.

But things were not so bad. Since the 1930s, Fritz Zwicky's pioneering observations of the motion of galaxies in clusters showed that there is much more matter in galaxies and in clusters of galaxies than meets the eye. More accurate measurements during the last two decades of the twentieth century led astronomers to conclude that dark matter can add up to about 30 percent of the critical density, closer to flatness but not there yet.

Of the several puzzling facts about dark matter, none is more puzzling than its very nature. What is it, exactly? We still don't know! We know it is out there, and we also know that most of it is not made up of ordinary stuff such as protons, neutrons, and electrons. It is quite ironic—and humbling—that much of the matter that fills the universe is unrelated to the matter we are made of. This is exciting news for particle physicists—all this mysterious matter out there, waiting to be discovered. Quite possibly, this exotic dark matter, which interacts with ordinary matter only gravitationally (or very weakly otherwise), is part of a unified theory, be it at the GUT level or even at the superstring level. Indeed, several unified theories predict the existence of particles with the right properties to behave like dark matter, such as massive neutrinos. But the case for the composition of dark matter is still wide open.[4]

It was at this juncture that the two schools of thought mentioned above, the purists and the gazers, clearly disagreed. The purists argued that the flatness problem, the existence of an old universe, was compelling enough for them to stick with a flat universe. The gazers argued instead that the observations supported an open universe, mostly at 10 or so percent of the critical density. This view was summarized by Peter Coles, of Queen Mary College, and George F. R. Ellis, of Cape Town University, in their book *Is the Universe Open or Closed?*, published in 1997: "Please be aware at the outset of our view that, ultimately, the question of [the universe's flatness] is an observational question and our theoretical prejudices must bow to empirical evidence."[5] This is, to say the least, a very sound scientific position. We may be inspired by our theoretical longings, but never to the point of blindness. As my grandfather used to say, a hat bigger than your head covers your eyes.

The purists maintained that inflation predicts a flat geometry for the universe. As the universe went through a brief period of accelerated expansion, whatever curvature it may have had was flattened away. The same happens to a balloon as it gets inflated. Focus on a patch on its surface; as the balloon grows, the initially curved patch becomes progressively flattened, until it is indistinguishable from a flat tabletop. Of course, the balloon as a whole is still curved, but the local curvature of the small patch is negligible. Inflation makes sure that the observable part of our universe fits well within a patch that was flattened almost to perfection; for all practical purposes, we cannot measure any curvature. (Note that the universe may still be curved at distances beyond what we could ever measure, that is, beyond our causal horizon.)

It is easy to understand why many cosmologists are so attracted to the inflationary universe idea; with one single sweep, it solves two key riddles of the standard big bang scenario—the horizon and the flatness problems. To see how inflation accomplishes this, we must investigate the mysterious mechanism that drives it: What can possibly make the early universe accelerate so substantially? First, recall that the expansion of the universe is an expansion of its geometry; picture two points on the surface of a balloon drifting away from each other as the balloon grows. Whatever drives inflation must stretch the geometry quite dramatically. We saw before that Einstein had proposed a sort of "antigravity" term in his equations, the cosmological constant, to counterbalance the implosive tendencies of his static universe. In other words, the balance between the attractive matter and the repulsive cosmological constant stabilized his static universe. Now, imagine a hypothetical universe devoid of any matter, whose evolution is completely determined by a cosmological constant. From the argument above, it is clear what will happen to its geometry: the repulsive cosmological constant, acting alone, will drive an accelerated expansion! This was already known in 1917, thanks to the work of Willem de Sitter. Of course, this universe is not our own, since we know that for most of its history the universe was not accelerating (moreover, it is certainly not empty). But, if we could arrange things so that something like a cosmological constant dominates the energy density for a short while and then goes away (by relaxing to zero), the universe would accelerate during this short phase and then go back to its normal decelerated expansion. This is precisely what happens in an inflationary cosmology.

The Cosmos, circa 1998: The Quest for New Cosmic Origins

The period of inflation was driven by what is called the *potential energy* of a hypothetical field, inspired originally by the Higgs field of the GUT model. In the dozens of versions of the inflationary idea proposed during the last two decades of the twentieth century, the driving mechanisms have changed, but the core idea of mimicking a cosmological constant for a brief period remained. We saw earlier that there are two kinds of energies in physics: energy of motion, the kinetic energy; and stored energy, the potential energy. An unrestrained system with potential energy will move, as potential energy is transformed into kinetic energy. If you raise a rock above the ground, it will have stored gravitational potential energy; if you let go of the rock, it will fall until it hits the zero of energy, in this case the ground. If you bring two electrically charged particles to within a certain distance from each other, they will have stored potential energy; if they have opposite charges and you let go, they will move toward each other, and if they have the same charge, they will move away from each other. The point is that if there is an interaction between objects (gravitational attraction between the rock and Earth, electrical force between two charged particles), there is the possibility of storing different amounts of potential energy. While there is a stored amount of potential energy, the system is not at its zero of energy and it will move (or try to) until it gets there, like the rock that will stop only once it hits the ground.

The field driving inflation—called the *inflaton*—can have two kinds of interactions: it can interact with itself via some force, as two electrons do via electromagnetism, and, since it has a mass (and energy), via gravity. (In some models, it can also interact with other kinds of particles, but the net result is essentially the same.) Different inflationary models assume different kinds of interactions, but they all work to provide the inflaton with some potential energy. The key point is that this potential energy has the same effect on the geometry of the universe as does a cosmological constant; it will drive an accelerated expansion, acting as a repulsive source of gravity. "Now, how can that be so?" you may justifiably ask. The explanation has two steps. First, in general relativity not only the energy density but also the pressure of matter contribute to the bending of the geometry. We can think of it as the energy stored in a tensed material such as a stretched rubber sheet. (For general relativity, things are more subtle, though. A stretched rubber sheet tends to go back to its original shape and not get stretched farther.) Second, the pressure of the inflaton's potential energy is negative. This negativity of the pressure will drive the accelerated expansion. If you bear with me for the next few lines, I will explain why.

An expanding normal gas has positive pressure; as the gas expands, it cools and its pressure drops. Likewise, a universe filled with hot gas decelerates and cools as it expands, since the gas pressure drops. In the case of the inflaton, it will be its stored potential energy and its associated negative pressure that will make the geometry accelerate. Just as the amount of gravitational potential energy stored in a rock grows as we increase its height, the amount of potential energy stored in the inflaton also grows as it takes values away from zero. (Think of the inflaton's value away from zero as its "height," like the rock's height from the ground.)

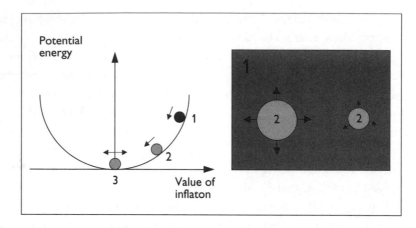

FIGURE 25: Left, *inflaton rolling down its potential energy.* Right, *regions where the inflaton has smaller values (denoted by 2) than its surroundings (denoted by 1) grow because of pressure difference: the pressure outside is smaller than inside the region.*

Suppose the inflaton permeates the early universe with a fairly smooth value. This will set its initial potential energy. Now imagine a tiny bubble within this "inflaton sea," within which the inflaton actually has a smaller value than outside, as is indicated in figure 25. This means that within this bubble the potential energy is smaller than outside. Now, in nature, systems left alone always tend to the point of lowest possible energy, just like the rock that falls to the ground. This means that the small bubble will tend to grow, lowering the total potential energy of the system. In order for the bubble to grow, it must exert pressure on its surroundings—that is, it must have higher pressure than its surroundings. So, we established that a bubble with lower potential energy has higher pressure than the surrounding inflaton sea. Taking this argument to the extreme, imagine a bubble with zero potential energy inside, indicated by the number 3 in the figure. Since this bubble is at the absolute minimum of the potential energy, it will also certainly grow. But since the inflaton within this bubble has zero value, the bubble has zero pressure! The bubble can grow only if its surroundings, the inflaton sea, have smaller pressure than zero, that is, negative pressure. Thus, any nonzero potential energy leads to negative pressure and to accelerated expansion.

Inflation ends when the inflaton rolls down to the zero of potential energy. The universe is a cold and empty place at this point, because any kind of matter or gas that may have been there before inflation has been diluted to almost total oblivion as a result of the rapid expansion. (If you "expand" a glass full of water to the size of a bathtub, the bathtub will be pretty empty.) Where, then, is all the matter that fills up the universe after inflation? This question is at the forefront of cosmological research. We believe that the inflaton doesn't simply freeze at the lowest point of potential energy but bounces around like a yo-yo (indicated by the double arrow in figure 25). As it does so, and because the inflaton interacts possibly with itself and other fields, a fantastic transmutation of potential and kinetic energy into material particles occurs, liberating an enormous amount of heat. In fact, within the inflationary scenario, we may legitimately call this period—known as reheating—the origin of the hot big bang model, an explosive creation of matter and heat.[6] Inflation reinvented the big bang. It is no coincidence that Alan Guth's popularization of inflationary cosmology is subtitled, *The Quest for a New Theory of Cosmic Origins.*

The Cosmos, circa 2002: Remembrance of Fluctuations Past

Inflation is a very compelling idea. All we have to do is assume there was an inflaton field early on in the universe slightly displaced from the minimum of its potential energy, and everything else follows: the horizon and flatness problems are solved, and even the matter that fills the universe is created during the dramatic finale to the accelerated expansion, as the inflaton yo-yos about the zero of potential energy. But this is not all inflation does. It also helps us understand a crucial aspect of the observations of the microwave background obtained by the COBE satellite and, more recently, by a host of other measurements. Recall that when I mentioned how extraordinarily smooth the microwave background was, I quoted its temperature as being homogeneous to within one part in a hundred thousand. This means that, in spite of its incredible smoothness, there are small fluctuations on the temperature of little over a millionth of a degree. To visualize how small these fluctuations are, imagine tiny ripples on the surface of a lake gently caressed by a breeze; if the lake is one hundred meters deep, the ripples are on the order of a tenth of a millimeter. There are high spots and low spots, which, for the microwave background, translate into hotter and colder spots in the microwave map. Two maps are shown in figure 37 in the insert, one by COBE, with an accuracy of about ten degrees, and one by the BOOMERANG experiment, with an accuracy of about one degree. Two new satellite missions will be collecting and analyzing high-precision data in the near future, the American MAP mission, launched during the summer of 2001, and the European Space Agency PLANCK mission, planned for 2007.

What mechanism caused these tiny temperature fluctuations in the microwave background? Whatever it was, it had to create inequalities in the gravitational pull acting on the photons; recall that strong gravitational fields cause photons to redshift, that is, become less energetic (colder) as they try to escape. We can picture the universe at about 300,000 years of age, when the photons decoupled from matter, as being permeated by overdense regions, where gravity was stronger than average. These regions would pull on the photons, causing the cold and the hot spots we see in the microwave background today; roughly, the cold spots are related to photons trying to escape the gravitational pull, while the hot spots are caused by photons falling into the overdense regions. These regions are the predecessors of the galaxies and the clusters of galaxies we see today, the seeds of the large-scale structure of the universe; once an overdense region forms, its gravity will induce ordinary and dark matter to clump around it. An extreme example is the so-called Great Attractor, a concentration of over 10^{16} solar masses, discovered in 1985 by a group of astronomers known as the seven samurai. Gary Wegner, one of my colleagues at Dartmouth (and one of the samurai), convened a conference here to discuss the observational results that led to the conclusion that such a concentration of mass exists.[7]

The standard big bang model does not provide any mechanism to create these overdense regions, having to assume a distribution of fluctuation sizes that matched astronomical observations, like a gardener who guesses what seeds were used in a garden by looking at the various blossoms. The inflationary universe, on the other hand, not only predicted the sizes of these fluctuations but also predicted them correctly! According to inflation, the shapes and forms of today's universe are direct descendants of tiny quantum fluctuations of the inflaton. It is a triumph of the inner space–outer space connection that the dynamics of the universe when it was only 10^{-36} seconds old can actually determine its large-scale properties today, the very small deeply interwoven with the very large.

The overdense regions that will affect the photons at decoupling are amplified—or better, inflated—quantum fluctuations of the inflaton field. According to quantum physics, everything fluctuates; you cannot pinpoint the exact position of an electron without incurring large errors in the measurement of its velocity. This somewhat paradoxical behavior is known as the uncertainty principle, proposed by the German physicist Werner Heisenberg in 1926. In other words, in the quantum world, nothing stands still. You can prepare a system at the minimum of its energy, sometimes called the vacuum of the system, but there will always be small fluctuations about it.[8] This is also true of the inflaton field; whatever its value and, thus, potential energy, there will always be fluctuations about it, little bubbles of varying sizes wherein the inflaton has a differ-

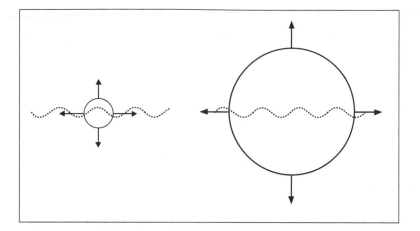

FIGURE 26: *After inflation, fluctuations* (dashed line) *are stretched to sizes larger than the observable universe* (left). *However, as the universe expands, these fluctuations reenter the observable universe* (right), *causing the overdense regions seen as temperature fluctuations in the cosmic microwave background. These regions are the seeds for the formation of large-scale structures in the universe.*

cnt value (see figure 25). During inflation, these tiny quantum fluctuations are stretched to enormous sizes, somewhat like the stretched bellows of a giant accordion. Some of them will be so huge that they will curve on length scales larger than the size of our observable universe for a long time, until we can catch up with them, so to speak (see figure 26). According to the inflationary cosmology, these amplified—or inflated—quantum fluctuations of the inflaton field will become, together with accreted matter and dark matter, the overdense regions that promote the hot and the cold spots in the microwave background observed by COBE and others. The discovery of these temperature fluctuations and the agreement with the prediction from inflationary cosmologies lent tremendous support to these theories. It is hard to think of alternative mechanisms that can do so much with so little.

The Cosmos, circa 2002:
The Return of the Ether?

There was, however, a very nagging problem. By early 1998, inflation predicted a flat universe and COBE seemed to strongly support that. But astronomical observations kept measuring the total amount of shining and dark matter to no more than 40 percent of the required critical density to explain the flatness. How could inflation be right in a universe so empty? Something was missing from this picture. In 1998, two independent groups of astronomers—one led by Saul Perlmutter at the Lawrence Berkeley National Laboratory, in California, and the other led by Brian Schmidt of Mount Stromlo and Siding Spring Observatories, in Australia, and Robert Kirchner of the Harvard-Smithsonian Center for Astrophysics—announced an amazing discovery: their observations of distant Type I supernovae indicated that the universe is expanding faster now than in the past. They found that these very distant objects, about half the age of the universe away in light-years, were receding slower than nearby objects. They concluded that when the universe was younger it expanded slower than it does now. Furthermore, and here is the clincher, their results implied that the universe is *accelerating* now! This news rocked the astronomical community worldwide. But there was more: not only is the universe accelerating now; it is doing so at a rate determined by a cosmological constant—or something mimicking it—which comprises 60 to 70 percent of the required critical energy density for flatness! The math is easy; add shining and dark matter at 30 or so percent of the critical density to this new repulsive gravity at 70 percent, and you obtain the needed critical density, reconciling inflationary theory and the COBE data with measurements of the energy density.

Because of their fundamental importance to cosmology, the observations of distant Type I supernovae have been subjected to intense scrutiny by the two groups and many others. So far, the results have survived. Effects that could mimic the extra dimming of the distant supernovae that led astronomers to their surprising conclusion, such as obscuring dust and variations on how early supernovae detonate, have not yet caused any major revisions in the data. However, given that we do not know much about how earlier Type I supernovae detonate, I believe we should take these observations seriously but with a grain of salt; it is always possible that supernova detonation in a younger universe was dimmer than in more recent times. Further studies, combining a satellite that will probe even deeper into the supernova population (called SNAP, Supernova Acceleration Project) with ground-based observations, will help clarify these issues.

Nevertheless, given the present robustness of these observations, cosmologists immediately responded by trying to make sense of this new and mysteriously convenient form of energy—dubbed dark energy—which dominates the expansion of the universe in an amount consistent with observations of the microwave background and the flatness predicted by inflation. Two questions spring to mind. What is this dark energy? And why has it started to dominate cosmic expansion only recently? We don't know the answers to either of these questions. As we have seen, the existence of a cosmological constant has been a very controversial topic in cosmology; it has been invoked in times of conflict between observations and theory, only to be summarily dismissed once these conflicts have been resolved. From the point of view of particle physics, the existence of a cosmological constant is a real problem, tangled up with the energy of the vacuum fluctuations mentioned above. Every matter or radiation field in nature should have fluctuations about its average value that, when added up, come to a truly enormous result, 10^{120} times larger than what is observed. Clearly, we don't know how to handle these vacuum fluctuations properly. The usual operational procedure is to sweep the dust under the carpet by saying that some as yet unknown mechanism operating at very small scales ensures that the vacuum fluctuations all add up to zero. Many proposals have been advanced to deal with this embarrassing issue, and some, such as super-string theories, are quite promising. But it is fair to say that the problem of vacuum fluctuations is one of the key open questions in high-energy physics.

The existence of a very small, but nonzero, cosmological term, as required by present observations, makes life even harder. It is much more satisfying to suppose that there must be a mechanism to *exactly* cancel a quantity than one that *almost exactly* cancels a quantity. A very elegant resolution of this issue, even if merely tentative at this point, is to suppose that the dark energy is due not to a cosmological constant term, the same across the whole universe, but to a field somewhat like the inflaton, displaced from its minimum of potential energy. Just as in inflation, if there is a bit of potential energy in this field, it will eventually drive an accelerated expansion of the cosmic geometry. (Recall that during the cosmic expansion, as matter gets more diluted, its contribution to the energy density gets progressively weaker. Thus, even though matter and radiation may dominate the early decelerating expansion of the universe, at some point the potential energy of this field will take over, starting the accelerated era.) The key difference between the inflaton and this new proposed field, called *quintessence*, lies in their relevant energy scales; while the inflaton dominated the expansion of the universe at its earliest moments of existence, at very high energies, the quintessence field started to dominate cosmic expansion only recently. This idea was refined in 1998 by my Dartmouth colleague Robert Caldwell, together with Rahul Dave and Steinhardt, and since then by many others. The key assumption is that the universe is permeated by an ethereal matter field (hence the quintessence denomination—the fifth essence of the Aristotelian cosmos) with a nonzero potential energy, whose main role is to drive its present accelerated expansion.

In spite of its attractiveness, the quintessence idea does not immediately answer the two questions raised by the observations—namely, what *is* the dark energy, and why is it important *now*. In regard to the first question, equating the dark energy with a dynamical field certainly addresses the issue of why there should be a small nonzero cosmological constant, as opposed to none at all. But it doesn't explain where the quintessence field comes from, that is, from what kind of particle physics model. This question, as I remarked above, is also unanswered for the inflaton field of inflationary models. The general expectation is that both fields (or could they be one field at different periods of activity?), if they indeed exist, are probably part of a more profound theory unifying the four fundamental forces. Whatever this theory may be, its predictions will have to agree with astronomical observations—producing, we hope, the two fields cosmology so badly needs.

The second question, why the dark energy started to dominate cosmic expansion only recently, is also a difficult one. Quite possibly, the resolution of this question is tied to the resolution of the first one; once we have a fundamental theory describing the properties of the inflaton and quintessence fields, we will know at what energy scales these fields dominate the cosmic dynamics. An alternative argument to explain the "Why now?" question, known as the anthropic principle, has the astronomical community deeply divided. The anthropic argument comes in two versions: weak and strong.[9] In its strong version, the principle states that the universe *must* have the right properties to permit intelligent life to develop at some stage in its history. Put another away, if there is a universe, there must be intelligent life. As the astrophysicist Mario Livio remarked in his book *The Accelerating Universe*, "The strong principle really crosses the borders of physics into teleology,"[10] by stating that the universe has as its purpose the creation of life. In doing so, the strong anthropic principle reverses the obvious historical trend of cosmology, which is to show how insignificant our existence is in the grand scheme of things; the strong principle puts us right back at the center. Furthermore, the principle is scientifically void, for it does not offer any mechanism to prove or disprove its foundational statement. The weak version states that the observed values of physical and cosmological quantities, such as the mass of the electron or the cosmological constant, are not accidental, but are determined by the requirement that carbon-based life forms (or possibly others) can exist and have already developed. In other words, the weak principle states that of all possible universes, young and old, and with different values of the fundamental parameters, this one surely matches the requirements for life; the universe is this way because we are here. Let's investigate this in more detail.

Assume that there is a multitude of universes out there—that we are part of a "multiverse"—where different universes (henceforth cosmoids) bubble in and out of existence for all eternity, some lasting for a very long time and others disappearing as fast as they appear. We can see how this idea emerges from inflationary cosmologies; as its main proponent, the American-based Russian physicist Andrei Linde, reasons, it is quite possible that different cosmoids within the multiverse will have different values of fundamental physical parameters, such as the cosmological constant or the mass of the electron. This being the case, some cosmoids will expand forever and produce life, others will quickly collapse, others may expand for a while, produce life, and then collapse, and so on. Of course, the multiverse hypothesis suffers from a serious problem: because these cosmoids are not causally connected to ours, we have not been and never will be able to communicate with them or even to test their existence. (There have been some speculations as to the possibility that information can be sent between cosmoids. But given the present uncertainties in the formulation of these ideas, it is wise to let them mature somewhat before publicizing them.) In any case, for a cosmoid branching out of the multiverse to be "successful"—that is, to live long enough to be able to support life—the fundamental physical and cosmological parameters must assume the values we measure in our universe. For example, a cosmoid with large values of the cosmological constant would expand too fast to form galaxies and thus stars and life, while cosmoids with a value even smaller than the current required value (a "vacuum" energy ten thousand times smaller than the energy binding a proton to an electron in a hydrogen atom) would be even more improbable than we are. We live in a very improbable cosmoid, but a successful one at that, capable of generating galaxies and intelligent life to ask questions about it. According to the weak anthropic principle, it is "clear" why the quintessence field started to accelerate the cosmic expansion only recently: had it been otherwise, the universe would have expanded too fast for galaxies to have formed, and we wouldn't be here.

I fall way to the extreme end of the group that deeply dislikes the anthropic principle, strong or weak. In my opinion, the principle lacks explanatory and predictive power, only justifying our presence in this universe a posteriori; it is the scientific equivalent of throwing in the towel, of giving up. The goal of science is not to justify why things are the way they are but to build testable explanations as to why they are the way they are. To say that we understand our presence in the universe by stating that we could live only in a universe old enough, and where the physical parameters have the values we measure, does not explain what physical mechanisms determined the universe's age and values of the physical parameters in it, but simply confirms that it could not have been otherwise. Perhaps the weak anthropic principle should be renamed the cosmological consistency principle, since it basically confirms that this universe is consistent with carbon-based life. Why? We still don't know.

I prefer to believe that our imagination and observations will keep on revealing the nature of the universe we live in. Our scientific explanations constitute a narrative, a representation of the physical world, as we are able to understand it through the use of tools, logic, and intuition. To marvel at cosmic coincidences, such as "if the electric charge of the electron was not exactly equal to that of the proton, then atoms would not be neutral and matter would not be stable," is to confuse our description of nature with nature itself. An electron exists only within the science we create; it is not a universal entity that we discovered. What we did do was interpret several observations through the creation of an entity to which we assigned the properties of the "thing" we call an electron. To put it differently, if there is another intelligent civilization somewhere in the universe, its members will also describe natural phenomena by means of their tools, logic, and intuition. They will create their own entities to make sense of their observations. Their science will be completely different from ours, even though both should be based on the same ultimate laws. We could compare their laws to ours and should agree to a proper translation. But they would not, in all likelihood, have created an "entity" such as our electron. When faced with the enormity of the cosmos, we should marvel not at how "finely tuned" it is for the existence of intelligent life, which implicitly asks for a teleological explanation, but at the fact that we have the imagination to comprehend and represent so much of it. We may be small players in the big scheme of things, isolated on a small planet, orbiting a small star among billions of others in an average galaxy, floating alongside billions of others in this vast universe. But our creativity spreads its wings across the cosmos, revealing worlds we may never touch. The humbler we are as we contemplate the unknown, the farther our flight will take us.

A Short Discourse on Time: Part 2

The possibility that we are living in an accelerated universe, driven by some cosmological constant or an ethereal quintessence field, forces us to revise the standard relationship between geometry and destiny. A universe with a cosmological constant does not obey the simple rules of Friedmann's cosmology we discussed in chapter 7, where the fate of the universe is uniquely determined by its energy density (see figure 19).

It is somewhat paradoxical that the more we know about the universe, the more we seem not to know. We do know that we live in a flat universe, and we know that its age is very close to fourteen billion years. For the universe to be flat, it must have three forms of contributions to its energy density: ordinary matter, like protons and electrons, which make up no more than 5 percent of the total; an unknown form (or forms) of dark matter, making up about 30 percent of the total; and an unknown form (or forms) of dark energy, making up the rest. We know the amounts that go into the cosmic recipe, but not most of the ingredients.

With a cosmological constant, it is possible for a closed universe to keep expanding forever, thus avoiding the "death by fire" of the big crunch predicted by Friedmann's cosmology. Of course, if we accept the present determination of the flatness of the universe, this exercise is somewhat academic; we "know" how large the cosmological constant must be. But, as we have learned from cosmology's history, it is always very wise to keep an open mind. The contribution from matter decreases with the expansion, whereas the cosmological constant remains, well, constant. Hence, in a closed universe, if the cosmological constant term becomes larger than the matter contribution before the turnaround point (see figure 19), when collapse begins, its repulsive gravity will drive an accelerated phase, avoiding the big crunch: in the presence of a cosmological constant, a closed geometry does not necessarily imply a big crunch. If a quintessence field permeates a closed universe, the situation is even more delicate. The quintessence field, acting just like a cosmological constant, may also come to dominate the expansion rate before collapse begins, thus avoiding the big crunch. But only temporarily. When the quintessence field relaxes to the zero of its potential energy, it will no longer drive an accelerated expansion. The matter density, albeit very diluted, will still have a chance to catch up and, eventually, reverse the trend causing the universe to collapse upon itself. Thus, it is really the nature of the dark energy that determines the fate of the universe.

Here things get somewhat complicated. We can always imagine, from the depths of our present ignorance, that the behavior of this dark energy may change in the future, say, by having a somewhat bizarre potential energy for the quintessence field, or different types of cosmological constants dominating at different times. In that case, unless we have complete knowledge of the properties of the cosmological constant or quintessence from a fundamental theory of the forces of nature, we cannot confidently predict the future behavior of the universe by simply measuring the present values of these parameters. As Krauss and Turner gloomily remarked, "We may never be confident that any presently inferred dynamical evolution can be extrapolated indefinitely into the future."[11]

If the existence of a cosmological constant or quintessence field is confirmed by future observations, we must abandon all hope of predicting the long-term behavior of the cosmos. Unless, of course, we do come across a fundamental theory of matter and forces, which will explain it all to us, from the nature of the inflaton field to the singularity inside black holes, and from the origin of the universe to the nature of quintessence. And why not call this theory GOD (Geometry of Destiny), as opposed to the current TOE (Theory of Everything)? After all, what separates us from the divine is our finite life span, the fact that our bodies perish in time, even though our ideas may remain. We exist within time, whereas God exists without. It is only through our creativity that we can transcend our corporeal boundaries and join the eternal. Our search for all-embracing meaning is no less passionate than that of Aristotle, Newton, or Einstein. A theory that can determine the origin and destiny of the universe represents our final rational embrace with the abstract concept of the divine, a culmination of a dream started with the Pythagoreans, who first believed that numbers described the essential nature of all things. Our yearning for a unified theory could not have evolved outside a religious culture deeply influenced by monotheism. This claim may cause some atheistic high-energy physicists to cringe, since they view the pursuit of knowledge as completely devoid of theological influences. In response, I ask them to consider the notion that we live within time and yearn to transcend it. The pursuit of an all-encompassing theory, rational and technical as it is, is also the passionate pursuit of something much larger than ourselves, something timeless, universal, all-determining. From it, we will obtain the "initial conditions" that set our classical universe in motion; from it, we will understand the behavior of matter and the fundamental forces at the smallest of scales, which we can then use to interpret the conditions inside a black hole or close to the big bang singularity; from it, we will uncover the fields that inflate the universe to flatness and that fill it up with ethereal energy; and, from it, we will determine the fate of the cosmos and, in doing so, our own. Even if we never achieve this lofty goal, we will live more fully for having tried. Our search is our redemption.

Epilogue:
Celestial Wisdom

When one tugs at a single thing in Nature,
He finds it hitched to the rest of the Universe.

— JOHN MUIR

The skies are full of magic. And this magic compels us to look up, to explain, somehow, our place in this vast cosmos. After all, we are stardust, the chemistry of our bodies derived from stellar explosions that occurred well before the formation of the solar system. If, during the history of humankind, our interpretations of celestial phenomena were originally put forward by the various religions, today they are derived from science. However, as I have tried to argue in this book, there isn't an abrupt rupture between the religious and the scientific discourses. The awe of and fascination with the skies and its mysteries, which are an integral part of most religions, influenced and still influence the development of the scientific theories created to explain the motions and properties of the celestial bodies. What was once unexpected and terrifying, so often interpreted as a message from the gods or even as a portent of impending doom, is now incorporated in our cosmic theories, which strive to explain the many celestial phenomena as natural consequences of causal relations between material objects. But the magic, even if now part of the scientific discourse, persists.

It is difficult to accept the idea that we are insignificant within the cosmos, that our existence, as individuals or as a species, has scant influence on the unfolding of the myriad creations and destructions occurring across the universe. Is it a perverse twist of creation that we can wonder and question only to know we will never have all the answers? How can we reconcile our ability to reflect about the world and ourselves with the fact that we are transient beings of limited knowledge and wisdom? Perhaps an answer lies in the reality that our lives *are* limited by space and by time. Without limits there is no desire. And without desire there is no creation. Like stars, which generate pressure to survive the crush of gravity, we create to survive the crush of time. A life spent worrying about what or whom we won't have a chance to love or understand would be wasted; it would be a life focused on loss and not on the balance dictated by the celestial wisdom.

We saw how regeneration springs forth from destruction; asteroids and comets have fallen on Earth, extinguishing countless species, but allowing for the appearance of countless others; stars are created from the remains of others, recycling matter across each of the billions of galaxies; even our universe—possibly one among many cosmoids—has a history, although we still don't know how it began and possibly will never know how it ends. These processes of regeneration do not belong exclusively to the skies; they happen every day all around us. Every tree that falls feeds the ground from which many others will grow; each human life may give birth to several others and inspire countless more. We are complex beings, capable of the most beautiful creations and the most horrendous crimes. Perhaps, by learning more about the world around us, we will be able to see beyond our own differences and to work together for the preservation of our planet and species. There is much wisdom to be found in the skies; but we must be eager to learn. The first step is easy; it requires only that we look around with respect, curiosity, humility, and admiration.

Notes

Chapter 1: The Skies Are Falling

1. Quoted in Carl Sagan and Ann Druyan, Comet (New York: Random House, 1985), p. 15.
2. Ibid.
3. Ibid., p. 20.
4. Peter B. Ellis, The Druids (Grand Rapids, Mich.: Eerdmans, 1994), p. 53.
5. Ibid., p. 56.
6. Ibid. Readers who are familiar with the brilliant comic-book series Asterix the Gaul would know of the confrontations between Celts and Romans and the roles of the Druids as shamans and makers of "magic potions" using the mistletoe. When I was a teenager, I too wondered what Getafix, the Druid, put in that superstrength potion he made; I can attest that several home-brewed experiments were all failures.
7. Ibid., p. 130.
8. John B. Noss, Man's Religions, 4th ed. (New York: Macmillan, 1969), p. 45.
9. I thank Susan Ackerman, of the Department of Religion at Dartmouth, for mentioning this connection. It should be noted that only chapters 7–12 in the Book of Daniel are dated at approximately 165 B.C.E. Chapters 1–6 were written earlier, probably between 350 and 250 B.C.E.

Chapter 2: Heaven's Alarm to the World

1. Increase Mather, Heavens Alarm to the World: A Sermon (Boston, 1681).
2. Increase Mather, Cometography: A Discourse concerning Comets (Boston, 1683), p. 2.
3. Ibid., p. 62.
4. Ibid., p. 63.
5. Donald K. Yeomans, Comets: A Chronological History of Observation, Science, Myth, and Folklore (New York: John Wiley, 1991), p. 106.
6. Ibid., p. 96.
7. Eugen Weber, Apocalypses, Prophecies, Cults, and Millennial Beliefs through the Ages (Cambridge: Harvard University Press, 1999), p. 88. It is interesting to note that, as late as 1997, in

the best-seller *The Bible Code,* Michael Drosnin affirms that the 304,805 Hebrew letters of the Torah (the first five books of the Old Testament) can function as an oracle, when arranged in different combinations.

8. John Maynard Keynes, quoted in Giorgio de Santillana and Hertha von Dechend, *Hamlet's Mill: An Essay on Myth and the Frame of Time* (Boston: Gambit, 1969), p. 25.

9. Norman Cohn, *The Pursuit of the Millennium,* 2d ed. (New York: Oxford University Press, 1970), pp. 25–26.

10. Augustine, quoted in Weber, *Apocalypses,* p. 46.

11. Saint Augustine, *The City of God* [bk. 20, sec. 24], Great Books of the Western World, ed. Mortimer J. Adler, vol. 16 (Chicago: Encyclopaedia Britannica, 1990), p. 625.

12. Virgil, *The Aeneid* [bk. 2, lines 694–701], Great Books of the Western World, ed. Mortimer J. Adler, vol. 12 (Chicago: Encyclopaedia Britannica, 1990), p. 116.

13. Quoted in Brian E. Daley, "Apocalypticism in Early Christian Theology," in *Encyclopedia of Apocalypticism,* ed. Bernard McGinn, vol. 2 (New York: Continuum, 1998), p. 25.

14. Bernard McGinn, in *Encyclopedia of Apocalypticism,* p. 74.

15. Quoted in G. G. Coulton, ed., *Life in the Middle Ages,* vol. 1 (Cambridge: Cambridge University Press, 1930), pp. 2–3.

16. Mather, *Cometography,* p. 62.

17. Gary W. Kronk, *Cometography: A Catalog of Comets* (Cambridge: Cambridge University Press, 1999).

18. Quoted in Cohn, *Pursuit of the Millennium,* p. 68.

19. Quoted in T. O. Wedel, *The Medieval Attitude towards Astrology, Particularly in England* (New York: Archon Books, 1968), pp. 91–92.

20. Quoted in Kronk, *Cometography,* p. 239.

21. Edgar Allan Poe, "The Masque of the Red Death," in *The Unabridged Edgar Allan Poe*, ed. Tam Mossman (Philadephia: Running Press, 1983), p. 744.

22. Chaucer, *The Canterbury Tales,* Great Books of the Western World, ed. Mortimer J. Adler, vol. 19 (Chicago: Encyclopaedia Britannica, 1990), p. 279.

23. William Shakespeare, "Sonnet 14," in Great Books of the Western World, ed. Mortimer J. Adler, vol. 25 (Chicago: Encyclopaedia Britannica, 1990), p. 588.

24. Quoted in Daniel Boorstin, *The Creators: A History of Heroes of the Imagination* (New York: Random House, 1992), p. 322.

25. Quoted in Sara Schechner Genuth, *Comets, Popular Culture, and the Birth of Modern Cosmology* (Princeton: Princeton University Press, 1997), p. 46.

26. Quoted in Weber, *Apocalypses,* p. 197.

CHAPTER 3: MAKING WORLDS

1. My main source for pre-Socratic philosophy is G. S. Kirk and J. E. Raven, *The Presocratic Philosophers* (Cambridge: Cambridge University Press, 1971). In this section, all text in italics is believed to be original.

2. Plato, *Timaeus,* Great Books of the Western World, ed. Mortimer J. Adler, vol. 6 (Chicago: Encyclopaedia Britannica, 1990), p. 444.

3. Lucretius, *The Nature of Things,* Great Books of the Western World, ed. Mortimer J. Adler, vol. 5 (Chicago: Encyclopaedia Britannica, 1990), p. 64.

4. Aristotle, *Meteorology* [7.10], Great Books of the Western World, ed. Mortimer J. Adler, vol. 7 (Chicago: Encyclopaedia Britannica, 1990), p. 450.

5. Ibid. [7.15], p. 450.

6. Ibid. [7.20], p. 450.

7. Ibid. [7.30], p. 451.

8. Quoted in Carl Sagan and Ann Druyan, *Comet* (New York: Random House, 1985), pp. 26–27.

9. Quoted in Sara Schechner Genuth, *Comets, Popular Culture, and the Birth of Modern Cosmology* (Princeton: Princeton University Press, 1997), p. 19.

10. Quoted ibid.

11. Quoted ibid., p. 21.

12. Quoted ibid., p. 39.

13. Clarisse Doris Hellman, *The Comet of 1577: Its Place in the History of Astronomy* (New York: Columbia University Press, 1944), p. 83. It is not clear whether Regiomontanus did in fact perform these observations. They were extracted from a treatise titled *De Cometis,* whose authorship is not clear. More details can be found in Jane L. Jervis, *Cometary Theory in Fifteenth-Century Europe* (Hingham, Mass.: Kluwer Academic Press, 1985). In any case, someone used Regiomontanus's parallax method to study the comet of 1472, a great step in the development of observational astronomy.

14. Quoted in Schechner Genuth, *Comets,* p. 45.

15. Quoted ibid.

16. Readers interested in a more detailed account of the Copernican revolution may consult my book *The Dancing Universe: From Creation Myths to the Big Bang* (New York: Dutton, 1997; Plume, 1998).

17. John Donne, "To the Countess of Huntingdon," in *The Complete English Poems* (New York: Penguin, 1986), p. 236.

18. Schechner Genuth, *Comets,* p. 102.

19. Ibid., p. 128.

20. Isaac Newton, *Mathematical Principles of Natural Philosophy,* Great Books of the Western World, ed. Mortimer J. Adler, vol. 32 (Chicago: Encyclopaedia Britannica, 1990), p. 368.

21. Ibid.

22. Isaac Newton, *Observations upon the Prophecies of Daniel and the Apocalypse of Saint John* (Cave Junction, Ore.: Institute of Science and Medicine, 1991), p. 251.

23. Ibid., pp. 251–52.

24. Quoted in Schechner Genuth, *Comets,* p. 151.

25. Quoted ibid., p. 163.

26. Ibid., p. 165.

27. Ibid.

28. Isaac Newton, "Four Letters to Richard Bentley," in Milton Munitz, ed., *Theories of the Universe: From Babylonian Myth to Modern Science* (Glencoe, Ill.: Free Press, 1957), p. 212.

29. Quoted in Stanley L. Jaki, *Planets and Planetarians: A History of Theories of the Origin of Planetary Systems* (New York: John Wiley, [1977]), p. 112.

30. Quoted in W. Hastie, ed. and trans., *Kant's Cosmogony* (Glasgow: James Maclehose, 1900), p. 24.

31. Ibid., p. 76.

32. Ibid., p. 154.

33. Quoted in Jaki, *Planets and Planetarians,* p. 128.

34. Quoted in Schechner Genuth, *Comets,* p. 210.

35. Ibid., p. 211.

36. Ibid.

Chapter 4: Impact!

1. Carolyn Sumners and Carlton Allen, *Cosmic Pinball: The Science of Comets, Meteors, and Asteroids* (New York: McGraw-Hill, 2000), p. 158.

2. Walter Alvarez, *T. Rex and the Crater of Doom* (Princeton: Princeton University Press, 1997).

3. Dale Russell and Wallace Tucker, "Supernovae and the Extinction of the Dinosaurs," *Nature* 229 (1971): 553–54.

4. See, e.g., Stephen Jay Gould, *Dinosaur in a Haystack: Reflections in Natural History* (New York: Harmony Books, 1995).

5. Luis Alvarez et al., "Extraterrestrial Cause for the Cretaceous-Tertiary Extinction," *Science* 208 (1980): 1095–108.

6. Carl Sagan, *The Demon-Haunted World: Science as a Candle in the Dark* (New York: Ballantine, 1997).

7. Sumners and Allen, *Cosmic Pinball,* p. 123. This is a good reference work for asteroids, comets, and collisions.

8. It is worthwhile reading Tom Gehrels, "Collisions with Comets and Asteroids," *Scientific American*, March 1996, p. 54.

9. Freeman Dyson, *Imagined Worlds* (Cambridge: Harvard University Press, 1997), pp. 79–81.

Chapter 5: Fire in the Sky

1. Several books on Sun worshiping are listed in the bibliography. See, e.g., those by William Tyler Olcott, James George Frazer, and Victor Wolfgang von Hagen, as well as that of John B. Noss.

2. Quoted in John B. Noss, *Man's Religions,* 4th ed. (New York: Macmillan, 1969), p. 321.

3. Joseph Campbell, *The Hero with a Thousand Faces* (Princeton: Princeton University Press, 1973), p. 213.

4. In my *Dancing Universe*, I offer a fairly detailed account of the historical evolution of quantum mechanics. Here, I will focus on those physical concepts relevant for our investigation of stellar astrophysics.

5. Outstanding accounts of the history and main ideas of nuclear fusion and fission can be found in the books by Richard Rhodes listed in the bibliography.

CHAPTER 6: COSMIC MAELSTROMS

1. Lord Byron, "Darkness," in Frank D. McConnell, ed., *Byron's Poetry* (New York: Norton, 1999), p. 20.

2. George Gamow, *A Star Called the Sun* (New York: Viking Press, 1964), p. 183.

3. Quoted in Kip S. Thorne, *Black Holes and Time Warps: Einstein's Outrageous Legacy* (New York: Norton, 1994), p. 142.

4. Ibid., p. 145.

5. Ibid., p. 160.

6. Readers interested in a nontechnical account of the special and general theories of relativity may consult my *Dancing Universe* and other references listed therein. Here I will focus on those concepts relevant for our investigation of black hole physics.

7. Giorgio de Santillana and Hertha von Dechend, *Hamlet's Mill: An Essay on Myth and the Frame of Time* (Boston: Gambit, 1969), p. 91.

8. Quoted ibid.

9. Ibid., chap. 17.

10. Jean Pierre Luminet, *Black Holes,* trans. Alison Bullough and Andrew King (Cambridge: Cambridge University Press, 1992), pp. 130–31.

CHAPTER 7: FIRE AND ICE

1. More details and a comprehensive bibliography on quantum mechanics and twentieth-century cosmology can be found in my *Dancing Universe*.

2. Arthur Eddington, *The Expanding Universe* (New York: Macmillan, 1933), p. 80.

3. Ibid., pp. 80–81.

4. For a detailed history of the discovery of the cosmic microwave background radiation, see, e.g., John C. Mather and J. Boslough, *The Very First Light: The True Inside Story of the Scientific Journey Back to the Dawn of the Universe* (New York: Basic Books, 1996), Joseph Silk, *The Big Bang: The Creation and Evolution of the Universe* (San Francisco: Freeman, 1980), and George Smoot and Keay Davidson, *Wrinkles in Time* (New York: William Morrow, 1993).

5. Saint Augustine, *The Confessions* [bk. 11, chaps. 12–13], Great Books of the Western World, ed. Mortimer J. Adler, vol. 16 (Chicago: Encyclopaedia Britannica, 1990), pp. 340–41.

6. Paul Davies, *The Last Three Minutes: Conjectures about the Ultimate Fate of the Universe* (New York: Basic Books, 1994), p. 9.

7. S. G. F. Brandon, "Time and the Destiny of Man," in J. T. Fraser, ed., *The Voices of Time: A Cooperative Survey of Man's Views of Time as Expressed by the Sciences and by the Humanities* (London: Penguin, 1968), p. 157.

Chapter 8: Time Regained

1. Percy Bysshe Shelley, *Hellas,* in Donald H. Reiman and Neil Fraistat, eds., *Shelley's Poetry and Prose* (New York: Norton, 2002), p. 438.

2. An excellent reference for geometric theories of unification, including superstring theories, is Brian Greene, *The Elegant Universe: Superstrings, Hidden Dimensions, and the Quest for the Ultimate Theory* (New York: Norton, 1999).

3. For an excellent nontechnical introduction to the ideas of inflationary cosmology, see Alan H. Guth, *The Inflationary Universe: The Quest for a New Theory of Cosmic Origins* (Reading, Mass.: Addison-Wesley, 1997).

4. For more about dark matter, see Marcia Bartusiak, *Through a Universe Darkly: A Cosmic Tale of Ancient Ethers, Dark Matter, and the Fate of the Universe* (New York: HarperCollins, 1993), and Lawrence Krauss, *Quintessence: The Mystery of Missing Mass in the Universe* (New York: Basic Books, 2000).

5. Peter Coles and George Ellis, *Is the Universe Open or Closed?* (Cambridge: Cambridge University Press, 1997), p. 19.

6. The word "reheating" is, in my view, a rather bad choice, since it is not necessary that the universe was actually hot before inflation.

7. For one version of the discovery of the Great Attractor, see Alan Dressler, *Voyage to the Great Attractor: Exploring Intergalactic Space* (New York: Knopf, 1994).

8. The interested reader should consult K. C. Cole's excellent *The Hole in the Universe: How Scientists Peered over the Edge of Emptiness and Found Everything* (New York: Harcourt, 2001), dedicated entirely to the vacuum in physics and cosmology.

9. More details can be found in John D. Barrow and Frank J. Tipler, *The Anthropic Cosmological Principle* (New York: Oxford University Press, 1986).

10. Mario Livio, *The Accelerating Universe: Infinite Expansion, the Cosmological Constant, and the Beauty of the Cosmos* (New York: John Wiley, 2000), p. 244.

11. L. Krauss and M. Turner, "Geometry and Destiny," *General Relativity and Gravitation* 31 (1999): 1453–59.

Bibliography

Adams, Fred, and Greg Laughlin. *The Five Ages of the Universe: Inside the Physics of Eternity.* New York: Free Press, 1999.

Ahearn, Edward J. *Visionary Fictions: Apocalyptic Writing from Blake to the Modern Age.* New Haven: Yale University Press, 1996.

Alvarez, Walter. *T. Rex and the Crater of Doom.* Princeton: Princeton University Press, 1997.

Armstrong, Karen. *A History of God: The 4000-Year Quest of Judaism, Christianity, and Islam.* New York: Knopf, 1993.

Barrow, John D., and Frank J. Tipler. *The Anthropic Cosmological Principle.* New York: Oxford University Press, 1986.

Bartusiak, Marcia. *Through a Universe Darkly: A Cosmic Tale of Ancient Ethers, Dark Matter, and the Fate of the Universe.* New York: HarperCollins, 1993.

Boorstin, Daniel J. *The Creators: A History of Heroes of the Imagination.* New York: Random House, 1992.

——. *The Discoverers: A History of Man's Search to Know Himself and His World.* New York: Vintage, 1985.

Campbell, Joseph. *The Hero with a Thousand Faces.* Princeton: Princeton University Press, 1973.

Chaisson, Eric, and Steve McMillan. *Astronomy Today.* 3d ed. Upper Saddle River, N.J.: Prentice Hall, 1999.

Close, Frank. *Apocalypse When?: Cosmic Catastrophe and the Fate of the Universe.* New York: William Morrow, 1988.

Cohn, Norman. *The Pursuit of the Millennium.* 2d ed. New York: Oxford University Press, 1970.

Cole, K. C. *The Hole in the Universe: How Scientists Peered over the Edge of Emptiness and Found Everything.* New York: Harcourt, 2001.

Coles, Peter, and George Ellis. *Is the Universe Open or Closed?* Cambridge: Cambridge University Press, 1997.

Coulton, G. G., ed. *Life in the Middle Ages.* Cambridge: Cambridge University Press, 1930.

Dauber, Philip M., and Richard A. Muller. *The Three Big Bangs: Comet Crashes, Exploding Stars, and the Creation of the Universe.* Reading, Mass.: Addison-Wesley, 1996.

Davies, Paul. *The Last Three Minutes: Conjectures about the Ultimate Fate of the Universe.* New York: Basic Books, 1994.

——. *About Time: Einstein's Unfinished Revolution.* New York: Simon and Schuster, 1995.

Donne, John. *The Complete English Poems.* Edited by A. J. Smith. New York: Penguin, 1986.

Dressler, Alan. *Voyage to the Great Attractor: Exploring Intergalactic Space.* New York: Knopf, 1994.

Dyson, Freeman. *Imagined Worlds.* Cambridge: Harvard University Press, 1997.

———. *The Sun, the Genome, and the Internet: Tools of Scientific Revolutions.* New York: Oxford University Press, 1999.

Eddington, Arthur. *The Expanding Universe.* New York: Macmillan, 1933.

Eliade, Mircea. *The Myth of the Eternal Return.* Translated by Willard R. Trask. New York: Pantheon Books, 1954.

Ellis, Peter B. *The Druids.* Grand Rapids, Mich.: Eerdmans, 1994.

Fraser, J. T., ed. *The Voices of Time: A Cooperative Survey of Man's Views of Time as Expressed by the Sciences and by the Humanities.* London: Penguin, 1968.

Frazer, James George. *The Golden Bough: A Study in Magic and Religion.* London: Macmillan, 1974.

Gamow, George. *A Star Called the Sun.* New York: Viking Press, 1964.

Gleiser, Marcelo. *The Dancing Universe: From Creation Myths to the Big Bang.* New York: Dutton, 1997; Plume, 1998.

Gould, Stephen Jay. *Dinosaur in a Haystack: Reflections in Natural History.* New York: Harmony Books, 1995.

Greene, Brian. *The Elegant Universe: Superstrings, Hidden Dimensions, and the Quest for the Ultimate Theory.* New York: Norton, 1999.

Guth, Alan H. *The Inflationary Universe: The Quest for a New Theory of Cosmic Origins.* Reading, Mass.: Addison-Wesley, 1997.

Halpern, Paul. *Countdown to Apocalypse: Asteroids, Tidal Waves, and the End of the World.* New York: Plenum Press, 1998.

Hastie, W., ed. and trans. *Kant's Cosmogony.* Glasgow: James Maclehose, 1900.

Hawking, Stephen W. *A Brief History of Time: From the Big Bang to Black Holes.* New York: Bantam Books, 1988.

Hellman, Clarisse Doris. *The Comet of 1577: Its Place in the History of Astronomy.* New York: Columbia University Press, 1944.

Jaki, Stanley L. *Planets and Planetarians: A History of Theories of the Origin of Planetary Systems.* New York: John Wiley, [1977].

Jervis, Jane L. *Cometary Theory in Fifteenth-Century Europe.* Hingham, Mass.: Kluwer Academic Press, 1985.

Kirk, G. S., and J. E. Raven. *The Presocratic Philosophers: A Critical History with a Selection of Texts.* Cambridge: Cambridge University Press, 1971.

Kolb, Rocky. *Blind Watchers of the Sky.* New York: Addison-Wesley, 1996.

Körtner, Ulrich H. J. *The End of the World: A Theological Interpretation.* Translated by Douglas W. Stott. Louisville: John Knox Press, 1995.

Krauss, Lawrence. *Quintessence: The Mystery of Missing Mass in the Universe.* New York: Basic Books, 2000.

Kronk, Gary W. *Cometography: A Catalog of Comets*. Cambridge: Cambridge University Press, 1999.

Livio, Mario. *The Accelerating Universe: Infinite Expansion, the Cosmological Constant, and the Beauty of the Cosmos*. New York: John Wiley, 2000.

Luminet, Jean Pierre. *Black Holes*. Translated by Alison Bullough and Andrew King. Cambridge: Cambridge University Press, 1992.

Mather, Increase. *Heavens Alarm to the World: A Sermon*. Boston, 1681.

———. *Cometography: A Discourse concerning Comets*. Boston, 1683.

Mather, John C., and J. Boslough. *The Very First Light: The True Inside Story of the Scientific Journey Back to the Dawn of the Universe*. New York: Basic Books, 1996.

McGinn, Bernard, ed. and trans. *Apocalyptic Spirituality: Treatises and Letters of Lactantius, Adso of Montier-en-Der, Joachim of Fiore, the Franciscan Spirituals, Savonarola*. New York: Paulist Press, 1979.

———, ed. *Encyclopedia of Apocalypticism*. Vol. 2. New York: Continuum, 1998.

Munitz, Milton K., ed. *Theories of the Universe: From Babylonian Myth to Modern Science*. Glencoe, Ill.: Free Press, 1957.

Newton, Isaac. *Observations upon the Prophecies of Daniel and the Apocalypse of Saint John*. Cave Junction, Ore.: Institute of Science and Medicine, 1991.

North, John. *The Norton History of Astronomy and Cosmology*. New York: Norton, 1995.

Noss, John B. *Man's Religions*. 4th ed. New York: Macmillan, 1969.

Novikov, Igor D. *The River of Time*. Translated by Vitaly Kisin. Cambridge: Cambridge University Press, 1998.

Olcott, William Tyler. *Sun Lore of All Ages: A Collection of Myths and Legends concerning the Sun and Its Worship*. New York: Putnam, 1914.

Poe, Edgar Allan. *The Unabridged Edgar Allan Poe*. Edited by Tam Mossman. Philadephia: Running Press, 1983.

Rhodes, Richard. *The Making of the Atomic Bomb*. New York: Simon and Schuster, 1986.

———. *Dark Sun: The Making of the Hydrogen Bomb*. New York: Simon and Schuster, 1995.

Sagan, Carl. *The Demon-Haunted World: Science as a Candle in the Dark*. New York: Ballantine, 1997.

Sagan, Carl, and Ann Druyan. *Comet*. New York: Random House, 1985.

Santillana, Giorgio de. *Reflections on Men and Ideas*. Cambridge: MIT Press, 1968.

Santillana, Giorgio de, and Hertha von Dechend. *Hamlet's Mill: An Essay on Myth and the Frame of Time*. Boston: Gambit, 1969.

Schechner Genuth, Sara. *Comets, Popular Culture, and the Birth of Modern Cosmology*. Princeton: Princeton University Press, 1997.

Schwartz, Hillel. *Century's End: An Orientation Manual toward the Year 2000*. New York: Doubleday, 1996.

Silk, Joseph. *The Big Bang: The Creation and Evolution of the Universe*. San Francisco: Freeman, 1980.

Smoot, George, and Keay Davidson. *Wrinkles in Time*. New York: William Morrow, 1993.

Sumners, Carolyn, and Carlton Allen. *Cosmic Pinball : The Science of Comets, Meteors, and Asteroids*. New York: McGraw-Hill, 2000.

Thorne, Kip S. *Black Holes and Time Warps: Einstein's Outrageous Legacy*. New York: Norton, 1994.

Von Hagen, Victor Wolfgang. *The Ancient Sun Kingdoms of the Americas: Aztec, Maya, Inca*. New York: World, 1961.

Weber, Eugen. *Apocalypses, Prophecies, Cults, and Millennial Beliefs through the Ages*. Cambridge: Harvard University Press, 1999.

Wedel, Theodore Otto. *The Medieval Attitude towards Astrology, Particularly in England*. New York: Archon Books, 1968.

Yeomans, Donald K. *Comets: A Chronological History of Observation, Science, Myth, and Folklore*. New York: John Wiley, 1991.

Index

Page numbers in *italics* refer to illustrations.